Artificial
Human Sensors
Science and Applications

Artificial
Human Sensors
Science and Applications

Peter Wide
Örebro University, Sweden

PAN STANFORD  PUBLISHING

Published by
Pan Stanford Publishing Pte. Ltd.
Penthouse Level, Suntec Tower 3
8 Temasek Boulevard
Singapore 038988

E-mail: editorial@panstanford.com
Web: www.panstanford.com

British Library Cataloguing-in-Publication Data
A catalogue record for this book is available from the British Library.

Artificial Human Sensors: Science and Applications

ISBN: 978-981-4241-58-8 (Hardcover)
ISBN: 978-981-4267-64-9 (eBook)

Printed in the USA

Foreword

In order to live and thrive in an often challenging world, humans have evolved to understand their environment and then conceptualise notions that would allow them to evaluate the impact of different natural phenomena as well as of their actions.

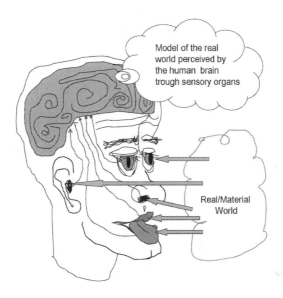

The need to develop better communications and trade relations through the history of humanity has led to a gradual development of the concept of *measurement* based on commonly agreed upon *basic units*, which allowed to assign symbolic and distinct "values" to different object parameters.

Vision, which is the most accurate and information-rich human sense, stood at the basis of the early measurement procedures for the evaluation of the *non-quantitative aspect attributes* such as colour, as well as the *quantitative geometric parameters*, such as length, surface and volume of the objects.

The evaluation of non-quantitative parameters such as colour is done by human decision makers who visually compare the specific colour against a reference set of standard-colour samples.

The evaluation of the quantitative parameters such as length, can be done either by counting how many *a priori* defined units have to be added together in order to best match the usually continuous parameter to be measured called the *measurand*, or by comparing the measurand against a specially made *graded scale*. In both cases it has been up to a human operator to look, compare and decide how many units the measurand was worth.

An ingenious solution for the measurement of non-geometric parameters such as time, weight, temperature, voltage, current, power, etc., is to convert those parameters into a proportional displacement of a pointer moving in front of a geometric scale conveniently graded in measurement units of the same nature as the measurand. Among these *non-geometric/displacement transducers* are the well known analog instruments: weight scales, clock watches, scale thermometers, electrical meters (for voltage, current, etc.), the oscilloscopes, magnetic and electric field meters, radioactivity meters, light intensity meters, etc.

It should be noted that *humans were essential integral components of the measuring process using these early meters*, as the measurement cannot actually be completed without having a person do the reading, i.e., visually deciding what numerical value should be assigned to the current position of the pointer on the graded scale.

The advent of the electronics technology in the 20th century allowed for the development of *automatic controllers* able to execute sophisticated control algorithms. As the "plain old meters" designed to provide data to human users could not satisfy the feed-back requirements of these controllers, a new generation of instruments had to be developed. These new *digital instruments* actually were the first *complete instruments* incorporating both the measurand-against-scale comparison and the generation of the numeric result.

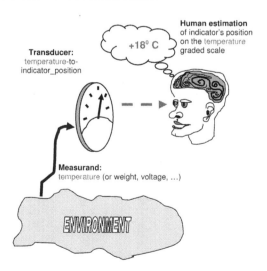

As an *effect of the digital instrumentation's success, the analog instruments became obsolete and humans were taken off the measurement loop* and relegated to global situation assessment and decision making roles.

Based on this digital instrument technology, new sensor systems have been developed for a large diversity of industrial (manufacturing, chemical, food production, pharmaceutical, etc), healthcare, and environment monitoring applications. Further developments in electronic and computer technology have allowed for complex data processing algorithms to be incorporated in intelligent *machine sensing and perception* systems able to explore a multitude of parameters over a broader frequency spectrum.

Recent developments in the computer and AI technologies have led to the apparition of a new sensing and perception paradigm, the *symbiotic human/instrument partnership.*

Humans and animals can act as sensor agents providing, usually fuzzy, *explicit estimates of specific parameters of interest* which they are naturally able to feel while artificial/machine sensors are not capable to measure.

Humans could be trained to estimate the value of the quantitative parameter of interest with varying quantisation errors and confidence levels such as *freezing, cold, cool,* or *around* $-20^{\circ}C$. Humans are able to recognise colours with a very high resolution. Dogs have been trained to recognise by smell even weak odour levels of substances like drugs or explosives, while pigs are used to detect truffles.

Observing non-verbal behaviour of humans, animals, or vegetation could provide *indirect-measurement clues about environmental parameters* such as ambient smell, radiation, air and water quality, extreme spectrum vibrations, etc., which are difficult or impossible to measure by instruments but are naturally detected by humans, animals, or vegetation, For example, canaries and mice were used for centuries as methane gas and carbon monoxide detectors in the coal mines to provide warning for explosion-potential and poisoned air. Leaf colouration and growth levels of plants and trees are used as qualitative indicators of environment status such as air and water pollution, temperature, etc.

Without entering into theoretical and technical details, this book represents a thought provoking introduction to the intriguing field of sensing and perception based on an ever evolving human-instrument partnership. It discusses new intelligent sensor technologies which are available today to enhance natural sensing, perception, and mobility abilities of humans allowing them to have a healthier, more productive, safer and overall better life.

Emil M. Petriu
12 February 2009

Preface

The human sensing capability is decreasing in capacity and affected by a general deterioration, mostly depending on generation heritage, civilisation, contamination, and in short range perhaps mostly affected by age. The primary motivation for human-based sensing or more correctly artificial perceptual sensor systems, therefore, is to explore an area of human complementary sensor systems that is foreseen to provide a new and emerging capability in human performance and well-being.

This book views an approach, that directs the possibility of increased perception, looking at the fact that we search for more experience and adventures in our lives. Also with a continuously growing population the scientific field of human-based sensing will substantially increase the interest for commercial solutions that complement the perceptual sensing ability of an individual.

Artificial-based sensing consider, for excellent reasons, a need for sensors and new sensor principles, in order to design systems that complement the human abilities for direct and indirect interaction with the real world, to get an increased quality of information. The interaction also provides challenges. The definition of the optimal communication flow — that is to receive the right information in time, with data that most likely is not redundant nor of good quality, also requires an ability to interact with the proximity of an individual human. The problem is that in most measurements there is a great demand for additional information in parameters that at a specific time is fragmentary and partially presented. If there are poor or insufficient data, there is always a possibility to access more parametric data or to find the requested information by complementary data. For example, by measuring, by an IR-sensor, the temperature in an ear of a child is a convenient technique to get the perception of sickness, that is the result provides a definitive temperature. This is done to estimate whether the child has fever, no fever or adjoin. However, this conclusion could also be estimated by complementary information, checking the overall condition of the child's behavior, red cheeks, crying, passive, sweating, freezing etc.

This perspective is actually the intention with writing this sensor book, to get an excellent possibility to present a deeper view, describing the importance of information to provide an apprehension and knowledge about sensors, their prospects and opportunities in general and human based sensing in particular. This is considered to be an area of highly interest for individual person's to get increased enrichment in life. This concept of sensor concept may also contribute to make individual's able to be more active also when the perceptual abilities decrease natural when getting older or by inability. The aspect of performing supporting systems to people who really needs them in daily life, adventure activities or just to perform better in sports, safety extremes as for example climbing in the mountains.

A strategy writing this book has constantly been to follow the intention to exclude all mathematical formulas and equations in order to focus on the primary concept as a textbook that easily can be read at the reader's convenience. However, a rich reference list referring to relevant complementary theories and complementary references for depth inquiry is given.

I hope that this book will be an inspiration source for anyone who has an interest in an emerging entrepreneurial domain and research field of great potential. Although, I foresee that students and researchers in different academic societies to learn more about the human based sensing and use this book as an inspiration source to further continue and develop the ideas and discussions. Also, other curious and practical entrepreneur-driven people is encourage to push this highly interesting area of technology based concept, and contribute to improve the human communication and interaction in order to achieve a valued social, active and dignified living for many people.

It is my clear view that the area of complementary artificial sensor systems that support, enrich and provide the interaction to a better life whether used for elderly, handicapped people, or just for a daily adventure experience, *will be one of the major research trends the coming decade.*

Indeed, the sensor abilities that reflect the real world and perform a vital function namely are providing a complement to the human perceptual system. This book is considered to be an inspiration guide that can be used by anyone to get an introduction into the new and emerging field of artificial perceptual sensing. It offers guidance on possible future developments without entering into a theoretical world of formulas and equations — there are plenty of references for the interested.

Editor

P. Wide

" ...a book for everyone and no one ... "

Nietzke

Contents

Chapter One

The Background

The word "sense" comprises the feeling of sensing the world. The perceptive view is complex, thus giving the user a remarkable possibility to register the close proximity in that part of the world they are acting and interacting with. The advantage of entering an expressive description of a new and emerging scientific field of artificial and perceptual sensor systems, is indeed a challenge that predicts an enormous scientific potential. Furthermore, to merge these systems to interact with an individual and its perception is an even more sophisticated task, at least if the ambition is to make a "true" symbiotic relationship between the system and its user. The general development strategy of a new technology system covering many traditional disciplines is naturally complicated to get a common understanding. Therefore, the requirements are focused on a convincing concept that has an illustrative view, which can be demonstrated through experiments and be easily described. The background application scenarios will hopefully view the benefits and show sufficient results in recognition of a number of situations described, that may be interpolated with the reader's own experience in similar situations. The merging of ideas given in this book, together with the reader's own experience, have hopefully the right mixture of innovative and constructive solutions that may forward the development of artificial human sensors in a new and unforeseen direction to the benefit of the user.

Human sensing is part of a perception process that is based on the fact that huge flow of information is constantly appearing in our proximity. In actual fact, most of the data is, however, not perceived and further considered as getting aware of it. We are living in a world with an overflow of information, of which a vast majority is considered to be unaware and therefore is not processed further.

To summarise this philosophical introduction, we are living in a world without much perception of what is happening around us, but we are still able to momentarily act on the sensations we are able to recognise and become aware of.

1.1 INTRODUCTION

The human perception is recognised as a well-adapted system that has emerged from the requirements and adaptations built upon generations of experience that have merged and tuned our sensing ability. As animals conducted specific sensing capabilities that increased the odds to survive and finding foods in a hostile environment, humans approached another direction by affiliating in social groups. Societies were built-up and organised in a constellation of civilisation, that in some sense formed rules. A main concern at that time was of course how to get the daily food. Society now has ways to prevent us from getting hungry by offering food in restaurants, supermarkets, etc., and when we experience danger, we have fire brigade, police, etc. Civilisation has also created a reverse effect on the human perceptual performance, by a slow but still noticeable degradation of the human sensing capacity.

With some gentle, but still provocative statements, we certainly can agree upon the following attitude:

Since generations in the past we do not, any longer, have the use for an optimal and development driven perceptual sensing performance.

The perceptual degradation is ongoing and we foretell the need of additional and complementary sensors to provide us with the right information at the right time.

We do not need more information but require right information to make use of and further develop our mental and physical strength as human beings because we always ensure that the quality of life is improved.

We actually do not emphasise to strengthen the ability to further develop our sensing capabilities and do not use the power of adaptation to the environment. This process has been declining for generations. The development is likely expected to continue, which can also be noticed in our daily lives. At the same time, we still suffer from the hereditary behaviours from an earlier time when we lived in cages, hunted animals and explored the eatable vegetation.

The forecasting premises to improve these conditions seems not to be too encouraging in the sense that our perception would probably continue to exhibit a decreasing development phase in the coming centuries. The reason is mainly related to the fact that the future need for skilled and developing perceptual abilities would most likely not be prioritised in the evolution process of man. The argumentation could of course find refutation of the assertions, but there has been an attempt to find external and logical references to academic work that supports this evolutionary phase, however without success. The reader only, can consider the credibility of this predication and is encouraged to find arguments to make an individual judgement. However, the arguments presented in this book are found in the reference list, that mainly consists of external arguments, basic data, tests and studies existing in the scientific literature.

Each of the following arguments is related to the basic human sensing ability and will describe a decadent trend in the perceptual performance as a further point to the discussion above. The arguments for the relation to the content of this book seems natural, that hopefully will evoke a constructive response from the reader, that there are general solutions available for artificial perceptual supporting systems. The arguments presented below have relevance in the research literature.

The timing is of course an important issue, due to the fact that it will always be presenting new and more thorough studies, the results of which will either strengthen or reject the academic arguments given in this section. However, as usual the readers are requested to make their own judgment in this matter and to keep in mind that the aim of this book is to motivate the need and emerging necessity of adding complementary sensor information to the human decision-making process.

1.2 THE ARGUMENTS

The examples in the following sections are meant to provide arguments for the fact that human capacity varies and their sensing ability differs quite significantly between individuals living in different environments and ethnical connections. This statement depends mainly on a number of reasons that create a more flexible living, where we are more or less able to take part in social society, adventure activities and enjoy life in a more pleasant way. The secondary effect that may be considered, is the additional aspect, that is affected by a large segment of the population that is getting older but is also more active, which makes us attain the flavour of an increased richness in our lives, as illustrated in Fig. 1.1. The figure illustrates the discriminating human being as a retired person with winter clothes on, looking through sunglasses, when eating a hamburger and listening to music with an Ipod.

The following four examples given in Fig. 1.1 are intended to demonstrate the wilfulness and obstinacy of the human social society, that is required in the accepted pattern of today's modern living. These social behaviours are taken without any consideration of its ancient origin, where the surviving concept was solely to live in connection with nature and interact with environment in such a way that we are considered to be a part of nature. Nowadays, it seems that the direction

Figure 1.1. The discriminating human being.

of man, instead, focuses on observing nature from the outside. The continuous feeling is that the existing requirements are not in balance with our environment. One reason for that can be that our sensing abilities are not adopted for the present, i.e. in the modern living pattern of today. The examples given below intend to show these differences and everyone may also calculate the effects that these limitations in perceptual performance may provide in the long term. However, by getting aware about perceptual limitations, we also are able to respond and either change our behaviours or adapt to the type of social life that is acceptable to provide the best possible condition for development.

1.2.1 The Skin Sense

The largest perceptual sensing organ is considered to be the skin of the body. The skin is the cover around the human that protects our internal organs from external damage. The perceptual organ, skin, is the interface through which we are able to touch the world and physically sense the environment.

Also in most cultures the skin exhibits a social identity by indicating the age, cultural affiliation, and also health status, social class or geographical connection of an individual. However, the skin is mainly the sensing organ, comprising of approximately two m^2 with an average weight of four kg (which naturally depends on the size of the person) and is constantly renewed throughout a person's lifetime.

Everyone may recognise their own perception of the skin when exposed to sun, heat, cold, pressure, etc. Indeed the skin is a major sensing organ that provides important information about the environment contact to the body. The human skin is, with some exceptions, in principle is hairless and this is connected to the body's ability to control the body temperature by the mechanism of sweating effectively and evaporation of sweat.

The influence of hair on the skin may affect the perceptual sensing capability of externally sensing the environment, i.e., the skin is able to sense the environment even if no contact with the skin occurs. However, most of the skin is directly connected to the perceptual sense of touch and has during evolution considered of being a central function primates explor the world.

The perceptual power of touch involves sensation through pressure, vibration or temperature in order to estimate the qualities, for example tactile texture. The tactile qualities will result in wide perceptual impressions ranging from sensual pleasure to extreme pain.

The skin comprises of a complex network of nerves, which form the connection between actual touch and the result experienced from that sensation. The nerve system is a connection of different receptor functions recognising the influence of contact properties. The tactile textures might be referred to (in contrast to the *visual* textures) immediate tangible skin contact, i.e., the feeling of an object surface. The tactile function has through generation back primarily been used for two main functions, to explore food and to communicate between humans. The interplay between senses is crucial to determine the quality of food or similarly when exploring behaviours. For example, when purchasing a peach in the food mart, the visual and olfaction senses are as important as to get the textural impression as touching the soft surface, maybe with the more sensitive outer part of the hand. Professionals often use the technique by using the outer and often hairy part of the hand, when detecting a presence of airflow, e.g., air flow leakage in a house. The same principle is used by the hair-bearing cells attached in the bat's wings, i.e., a function to detect the airflow on the wing surface. Otherwise, the advanced capability to externally recognise from, texture and temperature primarily by the hands is a helpful perceptual quality tool.

The role of the hair in relation to the skin sensing ability is obvious, but the hair can also be used in another dimension. The raising of the hair of the skin, i.e., piloerection is used in different occasions when feeling angry, frightened or thrilled and aim to communicate a state of looking larger and hopefully more threatening. Humans may in these situations feel this function when the hair rises on the arms.

The communication and social belongings by touch is a necessity, but in many cultures kept to a minimum, which may have negative effects. The touch between a mother and infant is well-known, generally recognised and accepted. However, the socially based skin contacts between older people often exhibit a completely lack of touch from other humans. The sensation of the skin has different zones of more or less dense nerve cells networks used to communicate their identity, social status and their sexual desirability. This is different in cultures where the skin is naked and shown to others. On the other hand, the people in geographically cooler areas are probably suffering from this perceptual ability, when most of the time they have to hide their skin from freezing by wearing warm clothes. This measure, which also prevents additional sun rays from reaching the skin, i.e., lack of the sun reaching the skin prevent important sun related vitamin D formation, as well as may cause depression. This phenomenon can be experienced during wintertime in areas where the absence of the sun is obvious during certain parts of the year, e.g., northern Europe.

Geographical factors can be predicted to have a major influence on the use of our biggest perceptual sensing organ, i.e., by sheltering the tactile sensing ability of the skin, will actually prevent whole populations to experience the full flavour of this sense. This restriction will most likely affect the tactile ability in a number of functions, social contacts, sexuality and emotions. In general, the restriction of using the skin capability in features like experience impressions from the environment, for example, electrical lightning, pressure change at a concert caused by the huge loud speakers or wind flow on top of the mountain. The prevention of environmental properties from contact with the skin will seriously affect the human perception and by that means also limit the power to receive optimal impressions.

It may not be too controversial to assume that the tactile perceptual organ is differently used in different climate zones and therefore will exhibit serious impacts on the perception in general and specifically the tactile sensing capability. What this means in the individual's personality, cultural or social affiliation is quite an interesting aspect, however not a question of this subject. Let us take note and recognise the fact that the available tactile performance is used differently depending on the environment conditions.

1.2.2 The Visual Acuity

The use of the vision system in different situations to discover, detect, track or follow the course of events, is considered of vital importance for man, and has always been a major quality in surviving a hostile world. In previous generations however, humans have entered into an evolutionary phase consisting of partly different challenges. The fact that we do not have to make use of the visual proficiency in our daily lives for survival noticeably increased in the last decades. The observation can mainly be related to the change of living conditions that pervades our lifestyle. The change in lifestyle has provided us with circumstances that are not needed anymore and we have to adapt to new conditions.

In the literature, a number of investigations can be found regarding adult population studies, showing that better visual acuity is generally confirmed at all ages and more so in men than in women, e.g., Robaei (2005). Also the age-related change in contrast is confirmed, e.g., in the Nomura (2003) study.

Among young people, a clear gender difference among children has not been firmly established in the literature, as reported by Robaei (2005). However, these results contrasts new findings revealing a significantly higher prevalence of visual impairment among girls in a Chinese study, He (2004). In contrast, the studies from South Africa, Naidoo (2003) demonstrated a higher prevalence among boys. This clearly confirms the need to conduct further research in this area that may clarify the differences.

This may, in fact, show other findings that are due to the greater variability of visual acuity gender and social activities between boys and girls. This indicates, among other issues, a socially different development phase in combination with geographical differences in populations between young men and young women.

The studies of ethnical and regional differences is also represented in the literature. The study from Merritt (1996) found for example, a lower prevalence of subnormal visual acuity in white Americans than in African Americans. Regional differences are indicated in a number of studies from different countries. For example, a low overall rate of visual impairment probably indicates a very low prevalence of myopia among six-year-old Australian children. The myopia prevalence can in the different regional differences vary as much as 20 % in similar age tested groups in Taiwan and Singapore, Robaei (2005).

Furthermore, these findings are according to the study consistent with the prevalence of myopia in Australian older adults, which also has been reported to be lower than among adults in Europe and the United States. Whether these regional differences are due to unique environmental influences remains to be determined. As usual, further studies into this area, could of course provide important information regarding the visual degradation in populations.

As argued, the presented research findings indicate a strong relationship exhibiting differences between regional studies in adult populations. By adding the obvious differences in ethnical and age related differences, this fact indicates that we will probably end up with a massive visual impairment of the population. However, in population studies between children or young adults there seems to be a difference between males and females in the literature, but more illustrative is an indication that this effect may also be affected by choice of education during the student period of life.

These findings can be argued to relate to the social habits of today. The use of near-vision abilities has dramatically increased at the sacrifice of the use of the vision sense for long distances. This is an obvious change in human perceptual degradation, that exhibits a long-term process along with the changing social trends in decreased amount of exacting activities, will force the process further. The young generation of today experiences a situation where individuals actually are living in a near-vision daily life. In a general aspect, a normal day of living in a big city may comprise many components of near-visual activities, for example reading in school, playing virtual games and watching television with many hundreds of channels, movies and music videos played in the mobile phones. The young generation does not in general and on a regularly basis use the long-distance vision capability. The scenario of future generations will probably be to cause the evolution to make the human adaptation to fade out the long distance capability and change the focus of vision to more active near-vision ability.

The visual impairment may be exhibiting an influencing factor in the increased aspect, when spending constantly more time in bounded environments, e.g., home, school, shopping centres, etc. Rose (2008) has presented a study that supports this association, namely to assess the state between the relationship of near, mid-working distance and outdoor activities with prevalence of myopia in school-aged children. The conclusion made was when more time is spent outdoors, other than for sports activities, was associated with less myopia. Further, Ip (2008) reported that in a population-based sample of 12-year-old Australian schoolchildren, myopia was not significantly associated with time spent in near work after adjustments of other factors, but there were significant independent associations reported with close reading distance and continuous reading. The conclusion made in the article suggested that these associations indicate that the intensity, rather than the total duration of near work, is an important factor. This may indicate that children often being outdoors, more frequently focus on long distant objects compared to their indoor friends may also use their perceptual visual sensing in a more balanced manner, when using the vision sensing system.

The conclusion made from these academic studies may indicate that the intensity of near work will increase myopia and visual acuity. However, the result may be more balanced and associated with increasing outdoor activities. Further, it can be found in the literature, indications that a population study shows that university students have an increased risk of myopia than non-students, Risovic (2008). Also in the study, Quek (2008), a 73.9 % prevalence of myopia was found in a group of Singapore teenagers. Possible reasons for this were related to current reading and writing habits, and reading at close distances in the educational system.

1.2.3 The Smell and Taste Ability

Chemical compounds in nature are features of importance and a survival source of information for animals. Tasting and smelling have been since ancient time essential influence in nutrition and food selection. The taste related gustatory and smell related olfaction systems are fundamental senses in the mating process for creating and maintaining territories. The olfaction and gustatory systems are in both cases referred to as chemical senses due to the fact that the adequate stimulus consists of molecules that bind to the receptors of the sensory cells. Animals use a process called chemo-reception when responding to chemicals in the environment.

The human chemical sensing system consists basically of two primary techniques to detect chemical presence.

> Firstly, by contact inside the mouth, which primarily relates to the four taste qualities which are sweet, sour, salty and bitter. Umami is also related to this group. The process provides a structural risk that possibly unhealthy compounds accidentally enter the body. The sensing system in this case is not effectively detected by the other perceptual organs, i.e., the object enters the mouth before a possible complementary chemical, visual, tactile or auditory evaluation has been made.
>
> Secondly, by distance in close proximity of the body, provides an effective selection of airborne chemicals for food and avoidance of potentially toxic compounds. This process is considered to be of benefit for survival purposes when making it possible for an evaluation that continuously can be performed outside the body and hopefully on a safe distance before the individual gets in contact with the chemical stimuli.

The perceptions of smell and taste are often combined, as interacting with food stimuli when swallowed or inhaled. The typical flavour is based on the individual conception concerning the object stimuli and experience. However, in spite of the fact that we all have individual preferences, there is a general understanding that the gustatory and olfactory systems are necessary to ensure a good quality of life. According to Spielman (1998), more than two million Americans suffer from disorders that affect the sense of olfaction and/or gustatory.

Regardless of the variety of the level of flavour preferences in a population, related to age, gender and social patterns, one important issue still arises, the question about what is considered to be "normal smell and taste performance" in a population.

According to Bear (1996), most of the chemical based flavour is collected by the olfaction system. Up to 75–80 % of everything we perceive, when having a nice meal is considered to originate from the smell and only a minor part can related to be received from the gustatory system. This may be of importance when considering that the elderly normally decline in their performance of smell capabilities but in general and according to the reference maintain their taste sensation ability. The other reflection made is that we probably should emphasize and rely more on the olfactory system when the food is still outside the body and can be rejected without any risk of poisoning the mouth. Of course the smell and taste sensations perceive in parallel when the food is placed in the mouth.

On the other hand, a part of the population is considered to be super-sensitive to certain chemical stimuli. This group of outstanding people is called supertasters, Bartoshuk (1996), and the supertasting group have significantly more taste buds in their mouth, approximately four times more than "normal" is not unusual. The fungiform papillae of supertasters are smaller and more densely packed. The pain fibres associated with the taste buds make the supertasters extremely responsive to the burning sensation of spices in the mouth. Women are more likely than men to be oversensitive to taste. About 25 % of the human population is extremely responsive to a variety of especially bitter and sweet compounds in food, and is referred to as the supertasters. On the other hand, another group of 25 % of the population, referred to as non-tasters, is relatively unresponsive to chemical compounds in the food. The genetic variation in appreciating the difference of taste occurs across the world and can be found to enjoy the flavour of spices, Bartoshuk (1993).

Nevertheless, regarding the generic flavour, approximately half of the population is, based on the above reasoning, considered to have responsive taste behaviours that vary within a range, from what we may consider as normal taste behaviour. The conclusion drawn from this reasoning is that we may consider a variation of taste in half of the population that is either sensitive or non-sensitive, to be able to enjoy the different flavours of a nice and seasoning flavour in the food. This fact may of course be a challenge for the chef in a restaurant, assuming that half the guests dining in the restaurant have various taste perception. On the other hand, could this fact result in the aspect that half the number of guests, i.e., the non-tasters and the supertasters, will not appreciate a tasty dinner menu? Following this reasoning, of course then the other half the number of guests will enjoy a tasty dinner menu. Indeed the effect of concentration on taste plays an important role. Also the taste interaction in foods is significant and has been reported by Mojet (2004) for young and elderly people.

This variety in human perceptual sensing is a complex pattern that is hard to measure, and also to describe in words, i.e., to communicate a sensation and feeling for the involved individual. Everyone, who asked a child how the food tasted, will probably get a single word response back - OK or AWFUL - explaining either of the two alteration levels "good-bad". There seems to be an easy explanation with a short answer, that communicates either of the two extremes, instead of explaining about nuances concerning the received sensations. The existing problem is exhibiting the lack of interactive capability when not being able to find words in describing flavours and impressions and further how to communicate the sensations to other peoples. This lack of descriptive properties, especially in the young generation, is a serious problem and will if continued impoverish the descriptive language of the flavour of sensation.

In some situations, the human senses of smell and taste in conjunction with other perceptual senses, e.g., touch, sight, and hearing are together a useful tool for measuring and experiencing the total concept of flavour. The capacity of the human sensory analysis is a complex process not completely known in detail. However, the human senses provide us with a possibility to describe the appearance, flavour and texture, as well as to experience the complex hedonic responses of impressions, e.g., in food products. Sensory analysis in its human based and instrumentation form has a long tradition in the food industry and is becoming an essential and valuable tool in product development, quality control and marketing aiming to increase the total flavour for specific target groups. For example, pregnancy affects the sensitivity for taste and smell and the women may experience different changes in sensitivity of related products due to a strong influence from smell and taste.

The extra sensitivity in perceptual sensing makes certain groups in our society extremely sensitive for sensing impressions, and where a majority of the population considers it normal, for example, the group of elderly people, Schiffman (1993). This may for the individual result in an unpleasant experience as well as create annoying situations. On the other hand, if a person is talented with an extra sensitive olfaction and gustatory sensing system (supertaste/smell) there is always a need for these specific perceptual performances in the fragrance related industry, e.g., testing whiskey, perfume or being part of a panel testing food products.

The perfume and fragrance industry has an important task in preventing us from smelling bad, as well as putting a specific odour to our personality. These arguments only testify the aesthetic importance that we consider odour to be in the social society of today. The odour seems to reveal an ancient need, as the odour extends the individual and subjective flagrancy of perceptual sensing with the obvious goal to add a specific personal smell to the body and to be identified with a flavour.

The economic value of the perfume and flagrancy industry is of huge importance. Only in Europe the sales of fragrances during the year 2005 have been estimated at 6.9 billion Euro, Loutfi (2006). The significance of human olfaction seems to be influenced by artificial fragrances and no longer exhibits the survival relation, where species would be non-existing in a world of danger, without the specific ability, and in every moment, to detect and recognise odours.

Despite the importance of the gustatory and olfaction system, humans still lack the proper vocabulary to describe the experienced odour or in general a sensation. Due to the present individual sensing mode or other psychological effects, odours are often described in vague terms, often varying in time with no or little relation to the individual's earlier experience and possible similar references to other odours, Dubois (2000). The olfaction system seems also to be involved in our interaction with other humans and is considered to be involved in the selection of a partner, Bear (1996). In the literature, animals use to relate smell in path finding, e.g., ants, Romanes (2008); sexual activities, Edwards (1990); and how insects are organised in a society, Gordon (1999).

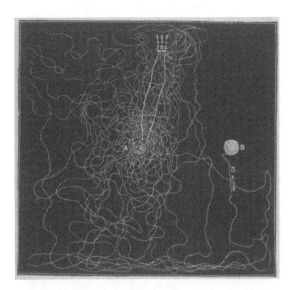

Figure 1.2. The pathways from Sir John Lubbock's scented experiment of ants finding their ways. Image first appeared in Animal Intelligence by George John Romanes (1881).

Smell is also considered to be of primary basic function for dangerous responses. A fundamental threat is alerting the basic olfaction functions. The result presented in Blanchard (2008) is considered to be clearly expected, and verifies as assumed the situational assessment when an individual is considered to be in a dangerous situation. The perception is then set to alert mode and the capabilities are focused to detect unforeseen activities in our close environment and to identify possible threats to us. There is also indication that we may learn to smell fear and danger by experience of earlier impressions with related olfaction experience.

Nowadays, smell is still used as a strategy of danger detection in emergency situations, where a suddenly arising gaseous compound is not expected to be present. For example, detection of smoke in the air or in a building will almost certainly put the brain into an alert mode. In the same way, we are also able to be alert when in the absence of certain flavours in the surrounding, for example, when having a dinner without recognition of the expected spices surrounding the specific food. The list of specific activities that with advantages can be identified by smell can be long. We indeed use the olfactory system in daily matters and as a direct primary attention, when we smell gas in or outside our car, when the cooking is going to be burned and when it is time to take out the garbage. Parents use with advantage this sensing ability to detect whether it is time to change diapers of the baby, as well as to make smell or taste checks when preparing food for young children.

The food industry incorporates frequently the use of sensory analysis, that has the final aim to develop products of the expected quality. Further, the effect is to understand the human food choice behaviour, where the human sensory analysis capability definitely can contribute to an overall impression. A complete overall impression should also include knowledge regarding food chemistry, physiological, psychological and sociological aspects. The ability to verbally be able to describe familiar food and liquid is of most essential importance in sensory analysis, as well as also to communicate the flavours to other people. The experience often obtained in these situations is that it may be easier to describe other perceptual senses, e.g., the appearance, sound and texture of a product than the actual sensation of a flavour. An essential area of interacting sensations, is with regard to whether the question arises if we are able to train to learn more, how to communicate flavour and be able to increase our capability to describe the experience of food. That would of course provide us with subjective benefits and the possibility to increase our individual interaction with the food store, when deciding to purchase. For example, apart from the colour of an apple, we will also receive a thorough description of the expectance in the richness of flavour when taking the first bite into the mouth. By being able to describe the flavour and the individual's taste experience of a food product, then it may provide the sensation from a delicious apple that is just appreciated by the expectations. Instead the usual procedure is, however, to randomly buy a red apple and hoping that the content, will taste according to the expected preferences.

The effects of a developed sensory language is mainly of advantage for vulnerable groups in the population, e.g., elderly people, on facilitating the description of smell and taste on nutrition and flavours, with regard to how food best suits the individual's attitude, preference and consumption. This effect may also be of importance when learning children to put words on their expectations and preferences as well as individuals who are, affected by medicine and diseases that reduce their sensations. By describing the needs to communicate flavour in foods according to individual demands, will most likely have a strong impact, and most probably contribute to the increase in quality of life, as well as the an interest for health-related consumption products, like for example vegetables.

1.2.4 *The Effect of Noise*

Sound is the sensation perceived by the sense of hearing and caused by mechanical radiant energy that is transmitted by longitudinal pressure, normally in air (Britannica). The impression of perceiving sound wave information is normally collected by the auditory system. However, when receiving the pressured energy in the air, by the body skin, an extended sensation may arise that will complement the complete sound perception. The total feeling of sensation of sound pressure can be experienced when the sound is reaching the body in an overwhelming contact with the sound energy, e.g., the sound from a thunderstorm. The auditory power is amplified through the body as providing the overall feeling, as for example, in an outdoor music concert. The perceptual sensing of sound is a continuously appearing phenomenon that on daily basis in a stream is delivered to us by the environment.

Music is a predetermined form of physical energy waves that is constantly present in our daily lives, and in its pure form is considered to establish a link between humans and the arts through our senses and enjoyment. This phenomenon normally creates a pleasure for the mind and may be healthy if conveyed by the body's own physical condition.

On the other hand, if the sound is experienced as undesired, unpleasant or unwanted, then it is defined as noise that may be considered as a personal health risk. Then in general terms, and due to this reasoning, the background sound in elevators, car radio, gym, shopping malls and at the dentists may be defined as background noise, i.e., unwanted sound. The background noise often causes communication problems to elderly people who have permanent hearing loss, when they desperately try to be a part of social life. Noise may also affect children and young adults where the risk of acquiring hearing loss has increased substantially in the last decades, Morata (2007).

A number of scientific publications have defined a risk moment in music induced noise that may cause permanent hearing loss, e.g., Kenna (2008). This auditory phenomenon is most likely based on the change in the young population's new behaviours and social pattern. The music induced noise behaviour supports the need of developing new techniques and provides alternative possibilities in designing new and better performing in personal music devices. The perception of ever increasing sound intensive levels during concerts and at night-clubs have made music exposure one of the most studied sources of excessive sound exposure to children and young adults.

New technologies have principally made the auditory and vision perceptual systems suitable for virtual enjoyments. Human perceptions, such as music, television, videos, games and other media have instantly made impressive experiences available for almost everyone. Since the introduction of personal cassette players in the 1970s, the development has been very technologically impressive and performance driven, aiming to deliver freedom of experience to each and everyone with an individual choice anywhere at anytime. Nowadays, we are able to listen to vast number of music materials for constantly long periods of time, and keep the equipment in our pocket or built in the mobile telephone. The compressed hand held unit has become popular due to their light weight and great versatility in storing different kinds of experience, making it easy to listen to music, play games or watch movies anywhere and practically anytime. The integrated music equipment has become a necessary icon for all ages, which we bring with us as a necessary tool, ensuring accessibility in our daily life.

The issue of personal music equipment has on the other hand brought up the question of potential risks to young people's hearing capability. This type of music overflow may not be considered to be noise, since noise by definition is unwanted sound, whereas music from a device is often quite the opposite. Hearing loss due to noise exposure may therefore be an individual health risk to anyone who is listening to abnormal sound. In several publications, a variety of implications have been made, such as portable music players may damage hearing, and also that there is a probability of MP3 players, and Ipods causing permanent hearing loss remains unclear.

The potential risks to children and young people with regard to hearing loss is not related to portable music equipment only, but has to take into account many sources in which youths may be exposed to loud noise. The term music-induced hearing loss is used for a condition similar to noise-induced hearing loss. Both noise- and music-induced hearing loss are characterised by a notch in the 4000 to 6000 Hz region of the audiogram, and are linked to hearing disorders, Morata (2007). The inference based on the given references may conclude that the total sum of both exposed music and noise will affect the risk of permanent hearing loss at young ages. The social pattern of today seems to increase the risk for young people to be exposed to increased sound that will provide a perceptual health problem. Fewer young individual's seem to have lost the preferences to experience a moment of silence in, for example, a secluded place in nature far away from the constant rattle and noise in a city, e.g., the communication systems like subways, trains, cars or air planes.

As hearing impairment among the young population rises due to mostly voluntary exposure to loud noise, there are many implications for health awareness. In the study related by Daniel (2007), the noise-induced hearing loss is reported to be a major cause of deafness and hearing impairment in the USA. It has been shown that though genetics and aging are major risk factors, temporary and permanent hearing impairments are becoming more common among young adults and children, especially for those with the increased exposure to portable music equipment.

But the music-induced noise is only one influencing factor of hearing loss. Another important factor of noise-induced hearing loss is related to education in schools. The classroom activities are especially vulnerable for human noise, when an individual cannot influence their choice of being in the disturbing environment many hours a day. Ristovska (2004), reported that school children exposed to elevated noise level had significantly decreased attention, social adaptability, and increased opposing behaviour in comparison with school children, who were not exposed to elevated noise levels. In a learning environment like a classroom, it is of most importance to provide adequate speech intelligibility to young individuals. This requirement is of vital significance, especially when the individual's learning process is affected and may in the worst case, lead to a reduction in learning efficiency. The classroom's noise level may affect the children's and student's ability to acquire basic knowledge and influence their future option for a qualitative good life.

As many studies have been conducted about specific noise source, such as classrooms, roads, building construction, the non-auditory consequences of typical and cumulated day-to-day noise exposure among children and young adults are poor. This cumulative effect includes both voluntary exposure music induced noise and as well as non-impressionable noise. The consequence is demonstrated in the literature, however, that the probability for many young children and adults is increased and will indeed cause an increased hearing risk, i.e., hearing loss and tinnitus. This argumentation will provide the implication that sound often virtually exposed in population, will provide a variety of hearing loss, i.e., a variation in the status of hearing in a population that will have restrictions on the expected social quality in life. Today, there is a tendency in the young generation to believe that a certain background noise is a normal state. The normal background noise has then no reference range, in the sense that no references have been experienced where the background noise effects are decreased to a minimum, or even disappeared in a quiet room, e.g., in a sound chamber.

1.3 THE AMUSED HUMAN

All animals need from their point of accessible capability to have fun and to curiously explore the adventured environment around them. The significance of human behaviour is however that there seems to be a divided situation, where a huge number of people concentrate their available time in survival activities, i.e., to get food for the day for themselves and their families. On the other hand, a part of the population experience their days by amusement activities, strongly encouraged by a healthy generation of aged people. However, the aged generation is in general equipped with limited perceptual abilities at least compared to the perception they experienced some years ago. The individual of today, with their limited perceptual utilisation, can still make use of the perception and experience to get amusements in excess of the ordinary working day.

The human sensing system is certainly a remarkable complex device, built up at our disposal to act as a window towards the world around us. Our sensing process can provide us with pleasure in life. In interaction with the collected sensations from the happenings, as we experience, it also makes us interfere with, and act in the dynamical world around us. This makes the human an independent, autonomous and interacting creature that appreciates to be an active part in the world. But a consequence thereof is also that the window towards the human proximity of sensing-acting around us is experienced differently by individuals, due to many reasons, for example, age, gender or cultural origin.

We enjoy sensory impressions, like listening to the high volume of sound from a rock concert, a vision inspired visit in a museum and the physical impression and enjoyment of walking in the forest. The feeling of being in the centre of a thunderstorm is an impressive adventure, the light enlightening the dark clouds, the rain rattling on our bare skin. Being part of a thunderstorm is not only perceived by the auditory sense, but the vibrating energy and light flashes of a thunder is perceived by our whole body in addition to experiencing the rain against the skin. This is indeed an experienced adventure that is perceived with the whole body. However, the experience from this type of overwhelming perceptual impressions is not a single experience, Our activities are accomplished by all possible and accessible perceptual abilities, e.g., the window towards the world is wide open that involves the whole arena of the human perception. We are not able to tune our senses or close the window to the world. Some perceptual senses are wide open, even in different degrees of unconsciousness. As a result of the shaping of the human perceptual senses, and in a degree of attention, we collect and deliver continuously a huge amount of information, even when we are sleeping. This amount of continuously received information has some connections and relations to earlier experiences and momentarily related input data.

Humans often correlate the sensing information with some earlier experienced situations and create a relationship to earlier episodes in life that is considered to be of similar interest and where knowledge may be drawn. The impression later in life, from these specific moments, may focus on some minor details that we remember clearly and relate to as a strong memory. For example, visiting relatives in the countryside when I was a child, was always a precious moment at each summer vacation. The details of how they lived, the surroundings and even how they looked like has since a long time faded away. But one specific feeling has remained in my memory and is clearly reviewed each time I recall these moments: – *the smell of the freshly baked buns.*

The afternoon coffee time was an event held in the garden, when the weather allowed it, and from these moments there obviously must have been a mix of incoming impressions from many perceptual sources. However, the impression of that specific olfaction sensation, accumulated during many summers when tasting these superb buns and receiving the smell from the buns, has certainly resulted in a permanent olfaction impression from decades ago. Today, I can still recall the flavour of the buns when relating my memory to these occasions. When thinking back to these summer vacations, the amazing feeling of enjoyment can be renewed in my memory and I can feel the sensation over and over again as illustrated in Fig. 1.3.

This overwhelming sensational feeling of that specific smell impression has in some sense faded away with age, even if the feeling in broad outline will remain in my memory. The olfaction sensation ability is most probably changing during life and time, and is most likely more sensitive at young ages when we are in an eager mode to learn and create a knowledge base involving all the new perceptual impression. When getting older, we have normally built up an impressive "register of experience" to use and compare similarities with freshly received impressions, that may compensate for lower olfaction abilities, e.g., aging. For example, when stating that it smells like old garbage, of course we need to have a clear picture and earlier experience of how old garbage smells. The brain processes the perceptual experience from earlier impressions caused by specific events and we are not able to consciously influence into the process, i.e., we may even not be aware of the process. By that means, we have no possibility to control or decide what momentarily sensing impression will be compared with in the brain. This phenomenon can also be experienced in persons who exhibit traumatic events. Actually, the perceived experience some time ago that is influencing present perceptions is indeed a specific eccentricity that makes a person individually unique and extremely complex,

Figure 1.3. The perception of newly-baked buns may give rise to long-lasting impressions. Photo courtesy and copyright Peter Wide © 2010.

when we also may need to communicate our decisions and maybe also the reason we acquired that specific determination with all its underlying elements.

We may also agree upon the statement that there is a substantial difference and a feeling of virtual situation in sitting on a chair and exploring an adventure maybe by supporting perceptual media in our own living room than experience the enjoyment in real life. However, how comfortable we may feel in a secure environment, by listening to our favourite music with accompanying video on a wide-screen plasma-tv of the latest model, with home surround sound system, the feeling of reality is often lacking.

A feeling occurs, that we are not able, even with all the accessibility to advanced techniques of today, and our perception cannot achieve the genuine sensation and the balance of reality of being there and experiencing it directly, is in some sense often missing. The feeling of special sensations by experiencing the wind on the face, standing on a high mountain, walking in a beautiful garden or listening to a rock concert gives unforgettable impressions. It seems that we have reached a level of "semi" virtual reality that is not designed to fully appreciate the human interaction. The human performance is simply not fully appreciating the level of interaction with existing techniques that are designed for perceiving today's virtual quality of life. This hereditary based characteristic is not able to create an optimal interface with today's existing technology platforms. This peculiarity can illustratively be shown by the miscommunication between separate senses that can be distorted. In that sense the individual is not able to make use of the information in an effective way. The data may appear in a time rate that is not appropriate for the human perception to receive, e.g., when driving a car. Further, data can be applied to the senses in such a way that even if we know that the human capability in this specific sensing mode is not perfect, we overestimate the perceiving data and miss important information, as when for example, driving in the dawn. There is a consensus in the literature that, the human hearing frequency spectrum whose performance generally decreases with, e.g., age and diseases. This quality is also valid for the olfactory sensing, which normally changes with age. Obviously age is one of a variety of important factors for exhibiting permanent decreased perception, which is also considered to be a major instant association, that excludes a well representation in a population. These people will most likely not be on an equal level of conditions in experiencing life. The emerging new advanced technology that is constantly introduced to the market does not, however, provide a perfectly adapted solution for each individual's specific discrepancies.

The virtual reality programs shown in computers, televisions or professional movies do not, for example, consider the degree of astigmatism that parts of the human population suffers from. The astigmatic error causes a feeling of seasickness appearing when interacting with a virtual situation and the experience of virtual beauty is therefore blown away. Vitale (2006), has conducted a study between 1999–2002 and presented that, refractive errors are common in the United States population. The study reported that more than 110 million Americans could or do achieve normal vision only with refractive correction. Obviously, there seems to be a number of individuals, who like me, are not able to enjoy the activities that can be found in computer simulators or three-dimensional movies.

The advanced technical systems for enjoyment that we experience today generally do not meet the expectations that a huge part of the population usually require. One main explanatory reason may be that the system normally does not interface with the requirements that our individual perceptual sensing system exhibit. In this book, there are also arguments provided that demonstrates different specific based unique and independent thresholds of the perceptual performance abilities, which make this perceptual capacity highly individual. To expand this reasoning, the dog of the family has a completely different experience in sensitivity and perceptual thresholds. The family animal has obviously another opinion of what is enjoyment in music or appetising odours in the environment than probably the rest of the family.

The revealing conclusion drawn from this exposition is that we probably do not comprehend the world in a similar way as other spices, or even between individuals. The fact may be annoying that there could be another world there outside, where animals experience another interpretation of the environment and explore in a different manner than humans. Other spices may then experience a more beautiful world than we ever can imagine by using other complex sensing abilities. Some animals have during generations refined the perceptual process, adapted and developed techniques to find an advanced and optimised performance in order to improve their sensation capability. Since the prerequisite of survival in the environment is crucial, then the condition and development phase

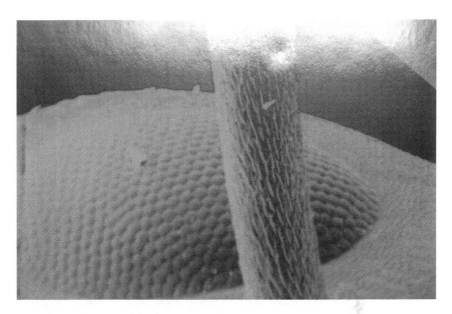

Figure 1.4. How will the world be perceived with this type of perceptual vision?.

normally differs, compared to humans where survival challenges are typically decreasing or have already disappeared in the daily life. The evolution process has been aimed to form the spices perceptual performance in a more goal driven and survival oriented direction, however during the last generations without too much of success. The example shown in Fig. 1.4 demonstrates the fact that the sensing input is quite different, more explicit and refined in different habitats, depending on their specific circumstances in the living environment.

Unfortunately, the unimaginable circumstances could be frightening, when we understand that we do not have a general view, or control of what may be happening in the world or even in our close proximity. Maybe Nietschke as illustrated by Heidegger, Gray (1968), has a crucial point, when stating that humans have no control of the world. Then, since we probably are not able to distinguish between the fine, decisive and crucial importance of perceptual sensations and specially developed nuance, as for example in olfaction in dogs, vision in birds or the tactile sensing on the skin of a shark.

Further, there could also be such an unrealistic circumstance, that we are not even able to perceive the full flavour of a complete world outside the one we are normally able to sense or feel.

Our human senses have since time generally decreased in performance and we may consider a need to again increase and complement the existing perceptual information to correspond to our actual needs and requirements. An increased amount of momentary information that complement our own senses could be of great satisfaction with regard to experience and provide an extended enjoyment, as well as a better quality of life. My spectacles (and the spectacles of my generation is indeed a proof of concept) is a good example that most of us need, or will need, to extend our vision capability by complementing with the addition of correction lenses to optimally adjust for an visual acuity. By complementing this simple correction to our existing vision system, we then are able, in a more active manner, to actively participate in the development in the environment and by that means successfully experience the world. This will, without doubt, increase the value of quality in life.

1.4 THE PRINCIPLE

The following principal illustrations in the next section may demonstrate the feasibility in a specific technical solution. The given example below will describe the principle in a complementing sensor system that is supported to the human perceptual system. A perceptual approach is shown to demonstrate the need for extended information, providing communication, and illustrated in a realistic application. The concept of artificial perceptual sensing is in this concept focused on the approach to find the coherent advantages that are complementary to human perception, and will provide a principal illustration of the area. The human perception has, in a historical perspective, been related to human safety aspects and often focused on the identification of hazardous situations or functioned as a warning system. This principal illustration could be a natural warning system in residents and effectively secure the water quality in your (and your family's) home.

1.4.1 A Principal Illustration

The human system has a remarkable ability to sense a dangerous situation in its close proximity, by using sophisticated perceptual qualities (e.g., smell, taste, vision, etc.). Also the ability to merge real time perceptual information with all the knowledge and experience a person may bring to the situation, will also characterise the human capacity. The possibility to complement the human perception with a remote network of perceptual-related sensors would naturally provide an added value of quality. However, an enriched information in exacting situations requires in principle that, in case of hazardous or attention-demanding situations in the proximity, a clear indication of the risk assessment is presented, preferably from a safe distance. That is, before the object constituting the risk gets in close contact with the human body, we need to get enough information to assess the situation. Therefore, a complementary and human-related perceptual ability, that provide correct, time dependent and added information about the situation of interest, is clearly an advantage. The design principle of a complementary sensor system is therefore of much importance, to work effectively in interacting between the system and humans. An effective interface should also consider that individuals with various degrees of sensing acuity will interact with this type of system. Therefore, a minimalistic, yet effective balanced information flow, should be favoured over

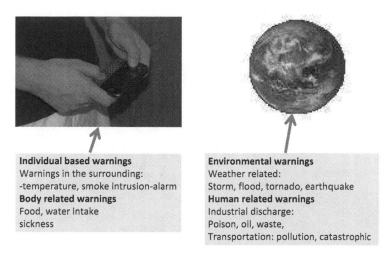

Individual based warnings	Environmental warnings
Warnings in the surrounding:	Weather related:
-temperature, smoke intrusion-alarm	Storm, flood, tornado, earthquake
Body related warnings	**Human related warnings**
Food, water intake	Industrial discharge:
sickness	Poison, oil, waste,
	Transportation: pollution, catastrophic

Figure 1.5. Example of individual safety system on the local (left) and global level (right).

detailed and overflow.

The proposed principle represents a new generation of sensors built for individual local use that can provide fast and accurate indications. The focus is on the goal to "extend" the human perceptual abilities by additional use of artificial sensors. With the word extend we also indicate a close connection between an individual and a sensor system that is able to interact in the proximity. There is a substantial benefit to displace the point of interest away from the human perceptual senses, e.g., indicating contaminated food before it enters the mouth would indeed provide a safer evaluation.

The emerging of sensor systems built for both individual (proximity) and global (remote) use can provide fast and accurate warnings to the individual level of safety. The warnings can be provided smoothly in close proximity of the body, e.g., in the clothes or the mobile phone. Attached to the warning, proper indication should also recommend actions to be taken on an individual level. In Fig.1.5, the principle of an individual safety and personal security approach using artificial and perceptual sensors can be provided in both local and global levels. As seen in recent catastrophes, we obviously need to extend our sensing ability by adding artificial sensors to our perception. This phenomenon is detectable by many animals, which obviously already have built in an extended warning system in their perception system by advanced perceptual sensing ability.

In addition to existing and large-scale measurement systems used today for general protection and warning to populations, e.g., tornado or earthquake monitoring, there is also a need for small sized monitoring on an individual basis. The example in the following section illustrates the human-based sensing principle, indicating a safety concept in close proximity to an individual, i.e., tap water quality evaluation. In Fig. 1.6, the illustration shows a normal state of drinking water, where the human perception normally indicates that the water quality is

Figure 1.6. An example of the human perception.

safe to consume, however often in a late state inside the body, i.e., in the mouth.

This state of behaviour is of course a normal mode of the individual, in the sense that the normal state is indeed fresh drinkable water. However, contaminated water may be mixed up with the fresh water and we have to handle that specific situation. In Fig. 1.7, an approach is illustrated, that indicates a complementary artificial sensor system that monitors the quality of drinking water. This technique is then considered as an additional complement to the human perception system. The individual does not need to bring the food inside its personal safety sphere or into the mouth before making the initial testing of the quality of the drinking water. Another motivation resides in a growing ability for artificial sensing systems to exceed the human perceptual performance in accuracy and precision in detecting contaminated water. For example, the human perception system is not always able to detect all chemicals or micro-organisms in drinking water by taste and smell alone. There is obviously a benefit to use an artificial sensor system that can find abnormal or unsafe compounds in the water and preferably before the individual puts it into the mouth. The outcome of this ability can be an increased interest of recognising chemical or biological compounds in the drinking water, where the individual interacts and makes prerequisite to learn to identify suspicious smell and taste flavours, i.e., learning to train the sensing abilities. The complementary sensor system will provide a conceptual solution that is intended to still maintain and preserve the area of individual integrity, as marked with three red lines in Fig. 1.7.

The communication between humans and a sensor system needs to use an interface that is easy to understand. A simple example could be a warning system inspired by the traffic light concept. A green light as seen in Fig. 1.7 corresponds to safe consumption and a red light indicates impurities of high concentration in the drinking water. Additionally, a zone of uncertainty may be added, represented as

Figure 1.7.   Example of a sensor system using an individual added monitoring system.

a yellow light. The yellow light corresponds to an ambiguity, i.e., cannot guarantee safe consumption, and further analysis may be needed. Representing the results as a traffic light makes it easier for the user to clearly understand the outcome of the sensor system and provide a safety concept.

In Fig. 1.8, the artificial sensing system complementing the human perception can be located at a distance from the individual, for example, at the water purification plant or even at the raw water source, i.e. the water reservoir, at a far distance from the individual. This concept illustrates a conceptual solution at both a local and remote view of the system principle. However, the overall aim in

Figure 1.8.   Example of a sensor system using remote monitoring in close proximity.

both cases is to present complementary information to the individual before any possible contamination got in contact with the body and may result in collateral damage. The distance examination of the most important food on earth should be in consideration to the diversity of the population. The variety of power of resistance does exhibit different levels of thresholds of contamination in the drinking water in a population, e.g., infants, patients and aged people.

The shown example illustrates a simple aspect, but however focuses on an emerging solution regarding an artificial sensor system for human assessment. The illustration is able to provide a safety concept in relation to an individual and is designed to operate in both proximity and remote levels. Indeed, this example would if applicable in almost every home, provide a safety concept for the whole family, by probable reduction of a major health problem or illness arising from contaminated drinking water.

As water is essential to life and health, it is therefore improbable that it is estimated that water affects one third of the world's population, by causing approximately two billion infections a year. According to (www.water.org), estimates that 320 million productive days in ages between 15–59 years, 272 million school attendance days, in addition to 1.5 billion healthy days for children under five years old illustrates only the seriousness to be taken, caused by contaminated drinking water (World Health Organization, WHO).

The use of artificial sensor system to prevent people from suffering from the most important, essential and basic food on earth — drinking water — seems therefore necessary. The effort to implement additional sensors that complement the human ability to survive and improve living conditions make for sure an augmented quality in life. This illustration is an excellent example of how artificial human sensors in the future may complement the individual's perception and hopefully be able to reduce these enormous numbers of illness that is caused by drinking water.

A question that arises concerns whether ancient man would have recognised chemical or biological contamination in greater extent than the man of today? Have the probable reduction of perceptual performance made us more vulnerable today than a couple of thousands (or even hundreds) of years ago? Are the obvious and present perceptual degradation problems of the population indirectly becoming a serious health problem?

The above questions, as illustrated in the example, indicate the quality of drinking water, certainly call for attention to the primary issue if there is a need for technical support systems. A complementary system that is in coherence with humans and interacts with the needs and expectations of a specific individual seems without doubts, an obvious and indeed an attractive solution. There are indications, however, that the emerging trends of future social behaviours will have the need for further extending the human capabilities. One of the major technical developments in the coming decades is foreseen to have a great demand for the direction toward artificial human supportive systems. These highly technical systems will effectively complement the individual based perceptual sensing ability and produce rich and timely information that hopefully will contribute to provide an enrichment for the individual's health and enjoyment.

The next chapter, intends to continue the initial introduction to the technical field and will as a concept present perceptual behaviours that are also considered to affect varieties of groups in a population. This perceptual quality may cause substantial differences appearing between individuals, which may lead to consequences for the individual's performance. For example, the colour vision will be discussed in more detail and have a quality variation that can affect the individual's quality of life. Indeed, the human population will be shown to exhibit a variety of perceptual performances.

The following chapters in this book will give further perspectives on artificial human based sensor systems, which are foreseen to gain the human sensing abilities, by technical solutions that will involve a multidisciplinary approach. Attached to each section are exemplified applications illustrating the field of interest and capabilities of human based artificial sensor performance.

1.5 APPLICATION

In daily life, we are continuously affected by circumstances that more or less have an influence on our perception, thus making us an active species in the world. This process is also influenced by whether we pay attention to the issue or are just passively floating on the information flow that is constantly entering our body.

Is the young generation of today able to experience an environment, free of disturbing noise and to rest the auditory system, by for example, entering a quiet room or sitting on a rock on the beach of a lake? This is indeed another experience, than the impression of sitting in a noiseless chamber, which will most likely establish an uncertainty and create a mixture of feelings. The human is through history closely connected to the nature and has since generation's back, been a part of and participating in, the evolution of nature. We usually feel more comfortable in a natural environment, e.g., the sound from water entering the beach than experience artificial sensations in a city environment or in a noise free chamber. It should be a human right for all children to experience these matters and feel the sensations from nature, without any disturbance from the artificial world, e.g., cities, train, cars, tv, etc.

Therefore, belonging to the young generation of today and growing up in a big city provide natural visual constraints. The daily living of the young generation, with activities like reading books, writing in school, game playing in free time, watching television and movies in the evening completes a day of near-visual perception. The concern is, however, that many young people of today exhibit the fact that they frequently experience, on a daily basis, a limited visual function. They are unfortunately not using their long-vision sight, which may result in consequences later in life, as well as for the future generations.

Another concern is that many people are raised in the tradition of fast food with no or little concern about the flavour and the process of food traditions, and especially appreciating the sensation of a nice dinner. The fact that many people are not able to describe the flavour of food or liquid is alarming and further, they are not able to communicate that impression to other individuals or receive other peoples sensational experiences.

The stress factor is considered to be one of the major issues that block the full possibility to obtain impressions from the surroundings and have the ability to focus on certain sensing impressions less possible.

This application section has the task to let the reader understand the underlying processes of focusing on themself and often subjective abilities regarding the process of perceptual awareness and attention. These simple tests can be executed at their best convenience and the user may reflect on the responses, based on their own individual status.

This type of perception oriented activities can be used to gain a relaxing concept and when stressing down also provide the possibility to experience the perceptual capabilities when, for example, focusing on the vision, smell, and hearing senses.

The following exercises may, however, be an illustrative example of getting to know your individual perceptual performance level and explore the specific sensing abilities. It has been argued earlier that there is specifically a rich variety in an individual's perceptual performance that depends on many circumstances, e.g., age, gender, social aspects and geographical causes.

An illustrative example of a relaxing exercise is derived from the Zen meditation and is described below. To be effective, this exercise requires at least 5 to 10 minutes and can be executed by old as well as young people.

Wear loose fitting clothing, with bare feet and stand in a, preferably outdoor environment as quietly as possible, look at a long-distant object and feel comfortable and unstressed.

Put your thumbs in front of you with straight arms and observe that you are able to use the long and short vision simultaneously by looking at your thumbs and a chosen long distance object. As focusing on the far object and the thumbs, you may discover the wide vision abilities by slowly, with stretched arms, moving the thumbs in a circular horizontal movement.

Start to slowly move forward in a suitable direction, continuously focusing on the far object. Put one foot in front of the other as shown in Fig. 1.9. First put the heel on the ground, simultaneously release the heel of the back foot. Slowly put the weight on the whole front foot and balance on the toes of the back foot. Then repeat the sequence by putting the next foot slowly in front of the first one using a speed of approximately one foot print per 30 seconds. During the exercise, focus on the far visual object and then put your hands in contact with your skin on the stomach, or simply put them together. Then consider the effect after the exercise, and experience the overall feeling from the different perceptual senses, e.g., the smell and hearing impressions. The time relaxing moment as illustrated in Fig. 1.10 can be executed in a varying outdoor location after a jogging session or

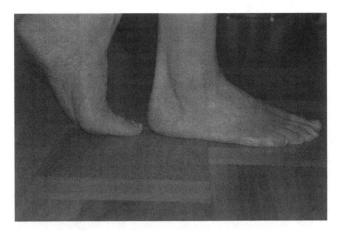

Figure 1.9. The walking schedule of the proposed relaxing exercise can be used by everyone. Photo courtesy and copyright Peter Wide © 2010.

Figure 1.10. A relaxing moment.

simply a stretching moment. The training activities should also be in parallel from the perception in receiving all the information from the surrounding.

The proposed illustrated exercise intends to exemplify the need for focusing on the individual internal perceptual performance and the existing single sensing abilities, before taking the next step and establishing a strategy of best using artificial additional sensor systems. For example, an almost blind person will most likely experience a marginal additional and perceptual effect when using a night vision system but may instead focus on developing the auditory sensory. The goal of this exercise is to make us more aware of the insight, regarding the capability of our human senses and how to experience the perceptual capabilities in an unstressful and harmonious environment.

Reports have frequently revealed that there is a dependency between the sensing modalities, indicating that the interaction between the human senses are not distinct, but rather cooperative. The sensory modality of vision affects the olfaction, which seems reasonable, e.g., if we hold a tasty cake in our hand, then we would have already noticed the visual input giving sensation of tastefulness. The same aspect would be valid if we are looking at a beautiful view, then it may easily be disturbed by an awful smell or unwanted noise. This indicate that the impression received by our sensory organs interact in a completeness that has to be taken into account. Therefore, when stressed or affected by external disturbances we are also affected by what we perceive.

Finally, to sum up this chapter the given examples indeed demonstrate that additional sensory information may, for many people, provide an increased complement to the existing perceptual senses. Principal technology approach examples will be demonstrated later in this book. However, artificial human sensors may be supportive if the individual can make use of the additional information, i.e., an effective interface is a crucial aspect in a symbiotic relationship between the user and system.

Chapter Two

The Context

The context of this book may be explained with the following sentences:

Is the mankind of today in her superior nature having control of her environment and does he actively interact in an optimal manner?

Is the nature of the mankind of today behaving in such a way that she succeeds to manage the environments, with the power given by the modern technology, and provide correct information to force mankind to make the right decisions? The unfamiliar decisions are often unpleasant but necessary for the survival of mankind as a superior creature.

The answer to these questions is most likely given by a simple **NO.**

Does mankind then need a supportive and artificial sensor system that will complement her perception to increase the possibilities to the right information given at the right time?

The answer to this question is probably a simple **YES.**

The context came up after intensive reading, sometimes also re-reading the book, referring to lecture notes, that was given almost 60 years ago in a course "What is called thinking" at the University of Freiburg, Germany.

(Martin Heidegger analysis of Nietzsche's book *Thus Spoke Zarathustra* (1883) in his university lectures course 1951/1952, *What is called thinking.*)

This type of intellectual thinking is of philosophical concern that should be recommended to each and every student. Especially, if they are interested in biological or artificial autonomous based systems.

2.1 INTRODUCTION

When Nietzsche's (1883) thoughts were formulated over 100 years ago, the author most likely did not even imagine the massive discussions that take place today would be centred around the measures taken and the efforts to adapt the human manner of living in a world of climate changes, war and hunger catastrophes. These natural phenomenon are still affecting the planet but people are more concerned about the former questions today. Could a reason for a more active individual participation to access the huge amount of updated information be available? Maybe the information is presented in a form and amount that it is hard for an

Artificial Human Sensors — Science and Applications by P. Wide
Copyright © 2012 by Pan Stanford Publishing Pte Ltd
www.panstanford.com
978-981-4241-58-8

individual to conceptualise and treat in a logical structure suitable for human pre-requisite.

The principal core point in this existential issue may lay in the following question:

Why we, as humans did not perceive the foreseen changes which have been going on for many decades?

Since the beginning of the industrial era, it has been a sign of success if the chimneys in factories produced as much smoke as possible. The huge amount of smoke indicated that production inside the industrial manufacturing plants was at a very high level and showed the surrounding that production was successful, as shown in Fig. 2.1. The cost of production, was at that early time of industrialisation relatively moderate in terms of manpower and energy. The industrial areas were however, heavily contaminated and as people moved from agricultural areas to cities and started to work in factories, a new era, the industrial revolution was born.

The human role in production was during the initial stages of the industrial era, where conditions were hard, dirty and in many cases affecting their health. Workers were relatively cheap to employ and the much manpower was used in the production process, the transportation industry, etc. The system was mainly built upon the human ability to work in different conditions and interacting continuously with industrial supportive technology. The technology development can be illustrated in, for example, the transport sector by giant machines like trucks, train and ships. The industrial era was in many senses a time period of global

Figure 2.1. The smoke indicates a state of success in the surrounding draw attention to mark a natural relation between production and chimney output. Photo courtesy and copyright Peter Wide © 2010.

growth and economic development, which also benefited of ordinary people. The human contribution was natural, as part of the industrial development and human skills were in many cases the reason for success in producing advanced products. The industrial performance in product and production advances was basically dependent on the workers skill and ability to use their perceptual power, the measures and values of the human senses. The skilled craftsmen were dependent on the human perception system to estimate the magnitude of qualitative characteristic, make probability judgement and use the flavour for fine-tuning a process. It should be mentioned that the industrial era in relation to the human senses was also an era of decadence due to the fact that people often worked in mainly unhealthy and contaminated environments, which often contributed to the degeneration of the capacity of the sensing capabilities.

Today, the industry is in an active process stage of development, where the industrial operations can make advanced operations without humans being involved. Commercial flights, trains and cars can be operated without humans in control, Urmson *et al.* (2006), Benjamin (2006), even if operational restrictions may prevent the use, Dalamagkidis (2008). Car bodies today are assembled in robot stations where teams of robots do their part in moving, positioning, gluing, welding and quality measuring in manufacturing. The manufacturing process requires few people in the production line to take care of the robots, i.e., to ensure that the robots complete their jobs within the expected precision and time.

The progress has in many cases moved further, that human skill is not needed in many advanced or unhealthy activities, where only some years ago, it was not possible without a human operator. In fact, the technology has in some cases been so safe, that humans now are the weakest link, Li (2002). Researchers estimate that approximately 80 % of all accidents are related to the human factor [Kirwan (1994)], and by studying the causes of near-accidents in industry, the ability for prevention has indeed increased, often by actively high-lighting the human influence in the process. Research activities in the area of increasing the interaction between individuals, computer systems and available resources of information are abundant and where it is applicable, aim to increase the symbiotic effects of the human involvement in advanced artificial systems. The goal is not to reduce human involvement in the process, but rather reduce the risk that human involvement introduces as an extra source of error in a symbiotic system. Also the use of robotic systems in contaminated working conditions provides us with an alternative not to use human perception abilities in dangerous situations.

In future, industrial work will most likely need to replace the human perceptual abilities and derive advantage from artificial systems that independently or in cooperation with human operators, e.g., in controlling a process or in building construction. The trend is however, that fewer people get more work done, by interacting with complex technical systems that will provide the main sensory information, dynamical information or logistic skills. The operator's main task will then in these situations be, restricted to control the artificial process operation, as in the example of the car body production described earlier.

This scenario will, as result in a course of events, change the era of industrialism and working conditions. A serious consequence may be that a large number of population that will be unemployed. To enable these people in old activities age to attend social activities and meaningful, the introduction of artificial sensors systems challenges the thresholds of perceptual performance that may shape enrichment activities and improve the quality life. Also, to focus on improving the unhealthy job conditions that is uncomfortable, dirty and unattractive works to be performed by mobile robot systems or automated operations as illustrated in Fig. 2.2. This may result in a social change and we may find a healthier population of all ages.

The reasoning given in the text above will give rise to the following two conceivable scenarios, that may require additional and sensing abilities to perceive additional information:

(1) The working part of a population has to increasingly spend more quality time in producing advanced services mainly in the maintenance and service sectors. A workers perceptual skills are nowadays in most cases replaced by corresponding artificial devices with better performance. The human perceptual ability provides a great possibility in creating new types of jobs in the service sector, providing a refocus of skill. This highly communicative skill is today, however not often required, when requiring an implementation of social, emotional based talent service oriented employment.

(2) The main populations, who will not work on a regular basis, also strive for a good quality in life. It is foreseen that this group is expecting to increase their

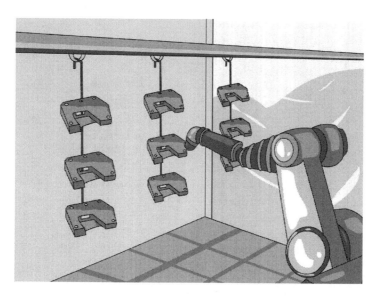

Figure 2.2. A robot substitutes the human worker in a contaminated environment, a painting cell.

level of enjoyments and experience, a new creating, growing and emerging society sector. By the increased number of active senior citizens, there will be a need for amusement and adventure specially directed towards the older age related segments.

When the available activities increase, and people have proper tools to experience adventure and advanced amusements, then a new industrial era of human needs will appear. This era of new discoveries is expected to bring people into new dimensions of life, for example, exploring space, experiencing high mountains and making excursions to the deep sea. Since we are not sure what kind of adventures would actually be appealing in a futuristic scenario but for sure the predictions are inviting and will indeed provide new perspectives in life.

The longing for complete, new experiences has roots in the human curiosity for discovering new objects and qualities in nature. This curiosity depends to a great extend the human need to excel and experience new sensations. We have a desire to explore the world around us and if the earth is becoming too small then we aim for the space.

This emerging area of technology-, social- and psychology-based skills with an increased use of information is expected to be an expanding field of interest in the coming decade. The possibility to increase human sensations is inspired and the outcome will hopefully create a richer and more active life on an individual basis.

To conclude this section, the question concerns the importance of human perception capability as a crucial part of the development of the world. Would people, as the great philosophers assumed, that mankind can be and is able to manage additional information from external sources, which are released as the nature of modern technology unfolds? Further, would an increased power of perception make it possible to come-up with better and well-reasoned decisions and directions for a structural development of the world? Would mankind with all its sorrow and joy be able to use an increased perceptual based intelligence to identify and realise unfavourable decisions?

I would, despite of the experience in history, relay the faith in a progressive and constructive technology development, as an emerging era of mankind and with a reasonable convincing argument state say a YES to the above questions.

The information concerning the example of the climate change was still in the early stages of the industrial era and also at the time of increased situational awareness already known. The Swedish researcher Svante Arrhenius received the Nobel Prize in 1903, for his discovery of climate changes. Maybe the time for perceptual awareness was not communicable into people's mind in the sense it is experienced today.

The sensation of ongoing climate changes was at least recognised by researchers. However, it is doubtful whether a person in a prominent position or any person could have influenced the economy-driven technology society to an extent, that the information would be taken seriously, even when the people could not perceive the changes in the climate.

2.1.1 The View of Artificial Sensing and Perception

The human perception can be viewed as a complex process, refined and developed since generations back, and exhibits several mechanisms for optimising the perceptual abilities to correspond to the expected needs. An example of information reduction is the handling procedure to manage the huge information flow that the brain continuously receives from different sensing systems. The ability to deal with a filter that make us aware of only the information we pay attention to and may have use of, is of importance for not overloading the perception process. That is, sensing information that we capture as an attention is evaluated with earlier experience and knowledge, but all other information is recalled as passive attention, and may pass the situational scene without having any impact on that specific information.

We may define two modes of perception, a passive and an active process, where a definition is made, as follows:

- The passive perception process refers to a passive attitude of an individual's behaviour, making no deliberate action or use of the continuously received sensing information.
- The active perception process refers to an active attitude of an individual's behaviour, making focussed and intentional actions and interacting with the outside world.

Actually, we make use and are attentionally aware of only a fraction of the incoming information that we constantly receive through our sensing mechanisms as information. In Goldstein (2006), the inattentional blindness is defined, where an article by a perceptual sensation that is not attended to is not perceived, even though a person is looking directly at that object. According to Norretranders (1998), we only actively process about 1% of the total incoming data.

The use of different sensing mechanisms, i.e., olfaction, vision, tactile, gustatory and auditory, interact and the received data are fused to a complete sensation. However, the influence of each sense is more differentiated and exhibits more importance to the human perception. As an example, vision is considered to dominate the human perception and figures of upto 90% of the total perceptual input has been recognised in the literature, Norrestranders (1998). Also, the relationship between different co-ordinating sensing processes differs, for example when tasting food, the flavour sensation collected from smell and taste are considered to be based on sensing information, mainly from the olfaction system, as much as up to 80%, Bear (1996).

It is indicated that people with sensing disabilities develop increased capability on the other senses, i.e. when they loose performance in one or more senses. Berlucci (2004), shows in a study that there is a new focus on the remaining sensing systems, performed by the brain in order to direct the information recourses from the lost sense towards other senses. In this study, a dissociation between taste and tactile extinction could be observed after a brain damage.

Individuals who are suffer from multiple sensing loss, e.g., a deaf-blind person or persons with similar experience by exhibiting deficiencies, can be compensated by an increased attention on the complemented perceptual organs. The redirection of focus may not substitute the previous attained information level, but give the individual a meaningful compensation for lost information. Regarding a deaf-blind individual, the most urgent sensation, according to a study presented by Ronnberg (2002), is to get early information of when, how, and from where other persons are approaching. Of importance was technical aids, which are dominated by visual effects, and vibratory senses were preferred. Complementary devices may be considered to provide support with existing solutions, like an electronic sensor stick, Sung (2001), with radio communication providing the individual an audio signal through earphones, Debnath (2001), or as in Batarseh (1997), producing a varying frequency of chirps that is inversely proportional to the distance measured.

There seems to exist a clear indication that individuals may compensate for lost information from one or several sensing organs by complementing with other similar sensing information in order to still make well, substantiated and structured ability and sensing performance. This decision-making strategy validates and makes decisions out of a huge dynamic data stream. The strategy is today, for example, used by business organisations in order to manage the huge amount of often fuzzy, unclear and sometimes "nonsensical" information that changes rapidly and can even be contradictory to each other. Also in military, rescue and health strategies, the use of right decisions made by as much information as possible is of vital interest to find the best momentary decision. The critical issue may often be to identify, evaluate and sort out irrelevant data (outliers) that could bias the decision and result in a non-optimal situation.

With an additional sensor information system that provides us with relevant complementation of the existing information and in a sort of loose connection, i.e., symbiosis with an individual, we have all the requirements to make more adequate decisions and facilitate the experience of a richer world.

The methodology of measurement capability with humans in symbiosis with an artificial system, is an emerging trend in measurement science. The methodology deals with the estimation of quantitative and qualitative parameters between human-computer symbiotic partnership, Petriu (2008). It exhibits a modified direction of symbiotic-human-machine technologies that has been used over the years for the development of more efficient computational intelligence and intelligent robot systems. However, this technology was originally aimed to be designed as human-based autonomous machines, and to be able to function in a balance of effective interaction with humans. The symbiotic interaction can be divided into two basic and principal groups (and one group in between) with regard to the system that interacts with humans. Additionally, a group in between, comprising a balanced mix of the two basic groups is essential to identify:

- the computer supporting system,
- the human supporting system and
- all the varieties in between.

In one basic aspect, the computer supportive artificial sensor system contains sensors that make the requested sensing and logical decisions, which are presented for the human operator in a complex system, e.g., in industrial, monitoring and surveillance operations. On the other hand, the human aspect can be used as a sensor node and make human based sensing. The sensing information is then communicated into the decision system, Pavlin (2007). The human supporting system approach makes the system redundant and very dynamic, knowing that people will report any unusual and threatening events to a computer system in a convincing and communicative way. This direction introduces a new methodology, where human perception is supportive as sensors in a complex system. Also a mixture of the basic directions above can be an attractive solution in designing a system's operational principle by using the best properties from both groups.

Following this direction of the discussion, then there is a predetermined structural pattern of how things around us behave in certain situations that seems understandable in how they act.

The observation of non-verbal behaviour of humans, animals, or even vegetation, could provide powerful and indirect measurement clues about environmental information.

Essential parameters that may provide useful information are, e.g., ambient smell, radiation, air and water contamination, and extreme spectrum vibrations. All of these could be difficult or hard to measure by instruments but are naturally detected by humans, animals, or vegetation. Historical examples of applications may, for example, be canaries and mice, which were used for centuries as methane gas and carbon monoxide detectors in coal mines to provide warnings for explosion and potential poisoned air, Petriu (2008). Rats and other animals are well-known to sense low frequency vibrations indicating earthquakes or volcano eruptions before these happen, Garces (2000). Also humans may be sensitive to sound waves, which they normally do not apprehend outside their hearing spectra. We are vulnerable to low frequency sound even if it is outside our hearing range. A feeling of discomfort, sorrow and fear appears, and evokes shiver when music is mixed with additional infra-sound. Leaf coloration and growth levels of plants and trees can be used, as qualitative indicators of the environment status, also additional qualitative parameters like air and water pollution levels, as well as temperature changes.

This sensing methodology indicates that there are additional aspects to consider, that can be used as advanced external structures for indirectly providing indicative and qualitative measures as a complement to humans and their limited perceptual performance.

Nowadays, nature-based biological indicators are being measured by additional new technology devices that have successfully increased the human assessment. The new emerging development of human self-diagnostics is an interesting future direction. New and specific sensors may be applied on the body, or built in the clothes, to detect, for example, if the human health status or its environment has changed. Sensors are then able to measure parameters such as blood pressure, heart and breathing activities and extreme loads of body parts. Also, the conditions of the interacting environment can be monitored in order to detect external situations such as unhealthy compounds, and humans can avoid threat-ening situations. One main question is how to solve the energy powering of the system, and the specific requirements to the sensors. These systems can accord-ing to Song (2006), be driven by converting mechanical movement energy, e.g., body movement, muscle stretching, blood pressure, vibration energy (e.g., acous-tic/ultrasonic wave), and hydraulic energy (e.g., flow of body fluid, blood flow, contraction of blood vessels), into electric energy. The energy generated may be sufficient for self-powering and low energy devices and systems in a variety of applications.

2.1.2 *Human Involvement*

The degree of human involvement in complex processes depends on the variety of the human operator's perceptual ability, experience and earlier knowledge, i.e., understanding of the systems dynamics behaviour. These facts cause uncertainty and make the symbiotic system reliable only when a rigorous control of each part's specific perceptual ability is known and well-defined behaviours are expected in different situations. With this follows the fact that the performance of each sys-tem is hard to evaluate when it is expected to be in a joint relationship with other biological/artificial systems. The involved systems also have to be effectively in-terfaced to each other in order to communicate properly.

Complexity differs in the context of human involvement and may vary between different operators that interact with the process. Figure 2.3 below, in-dicates the complexity in system configuration when interfering with the human

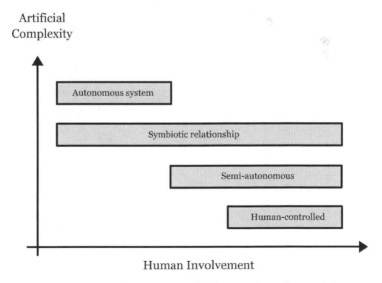

Figure 2.3. The complexity of an acting artificial system together with human involve-ments in a system with a degree of symbiosis.

skill. The interacting structure has to be considered as a two-way communication. The human symbiosis can be designed as a degree of acting in the system. The human participation may vary from measuring data to act as a network manager. The human may also be the acting decision system, receiving essential information in order to validate and make proper decisions.

Symbiotic human-machine technologies have originally been used over the years for the development of more efficient computational intelligence and intelligent robot systems. However, all these systems are designed as human behaviour based autonomous machines;

- autonomous, without human intervention, or
- semi autonomous, where the human is a predetermined active part of the system organisation.
- human controlled, with the on-line human operator providing high-level supervisory functions, Gill (1996),
- humans contributing with human-specific capabilities complementing those of the computer, Anderson (2003), in a symbiotic relationship in which each partner will lead in some cases and provide assistance in others.

The most challenging scenario is undoubtedly the leader/assistant role of a partner that will momentarily be decided on the basis of maximising the overall efficiency of the symbiotic team. Human beings are valuable in this symbiotic partnership to the degree that their capabilities may be favourable to complement those of the computers. Humans are still far more flexible and generally more intelligent than any computer, and are able to act on incomplete or ambiguous situations, as well as adapt to a variety of artificial interfaces. Humans have an advantage compared to artificial systems because they are able to interact directly with other humans unless disturbed by stress, tiredness or any other condition that may occur.

2.1.3 Context Awareness

The term human context aware in human-computer interaction means to react appropriately to the momentary circumstances that occur frequently when extensively and implicitly used — situational or context information — in interacting with humans. The computational context aware systems, Schilit (1994), refers to computing modes that are able to discover and react to changes in the environment they are situated within. The context aware human computer interface may increase the understanding of the reactive environment in order to achieve a higher abstraction level, aiming for a more effective human interaction.

Opposite to the research area of context aware human computer interaction is the natural relation to the human awareness expanding the model of being aware of the reactive world. This includes the interaction with a processing and valuation unit, e.g., the human brain. The human interfacing is a crucial step and one of the most fundamental tasks in building an effective and rational human interacting system.

The context aware systems have to make use of a new paradigm. The approach has to understand the context and relate to an interfaced sensor system that provides the right information at the right time when additional information is requested. Since humans usually find their own information, most suitably based on their own perceptual ability and merge it with their own huge brain related data base, it is of most importance to provide the corresponding artificial based information in association with the human mode of complementary interaction. This is valid when strengthening or weakening basic perceptual data by additional artificial data or in situations, when new and important data is presented, that sometimes may be out of range for the human organs.

Sensors systems based on the context aware principle will make the information available and aware to the human consciousness at a convenient time, however before a final decision is made, and when not distracting the human reasoning. Also, the meaning of interaction gives the possibility to provide directives and additional requested sensing to the context aware sensor system by showing abilities through training and teaching methods, Callinon (2007), or a strategic approach to plan recognition in semi-autonomous applications, e.g., wheelchair driving, Demeester (2008).

These qualities may also be used in an unconscious mode when the context aware-based sensor system recognises new sensing information of importance to the situational awareness. These situations may, for example, be if a complementary artificial olfaction system, i.e., an electronic nose system, identifies dangerous compounds in a close proximity that the human olfaction organ is not able to detect, e.g. carbon monoxide (CO). Since we cannot smell the CO, we are not able to feel a taste sensation that will make any impact on the context awareness. The artificial system is directly complementary in interacting with the environment rather than sensing the same properties as the human perceptions as illustrated in Fig. 2.4.

The symbiosis approach between human perception and artificial sensor systems provide a complex information interaction and deliberately makes distinction between the artificial and human capabilities. Further, it directs the sensing abili-

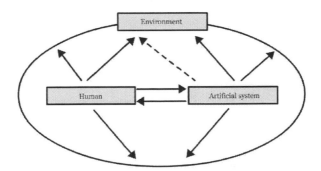

Figure 2.4. The conceptual Context Aware Human Machine Interaction, illustrating the complementary as well as supportive (dotted arrow) strategic information.

ties and performances of different approaches of perceptual sensing system. The human perception is a natural process of interaction including the brain. More detailed information and facts regarding the human perception process is available in the literature, e.g., Wenzel (1973).

2.2   LIGHT AND COLOURS

The human eye is sensitive to a narrow band of electromagnetic radiation that is available in the wavelength range between 400 and 700 nanometers. This bandwidth is commonly known as the visible light spectrum. The visible spectrum is located between the infrared and ultraviolet light, as can be seen in Fig. 2.5, which is the only source of detectable colours of the human vision. When combined, all of the wavelengths presented in the visible light, is about one third of the total spectral distribution that successfully passes through the Earth's atmosphere. The light forms colourless white light that can be refracted and dispersed into its component colours by means of a prism. The visible colours and subdivided into seven basic colours — red, orange, yellow, green, blue, indigo, and violet. The colours red, green, and blue are classically considered primary colours, because they are fundamental to human vision. Humans perceive light as white when all three cone cell types in the eye are simultaneously stimulated by equal amounts of red, green, and blue light.

A majority of the common natural and artificial light sources emit a broad range of wavelengths that cover the entire visible light spectrum, with some extending into the ultraviolet and infrared regions as well. For simple lighting applications, such as interior room lights, flashlights, spot and automobile headlights, and a host of other consumer business, and technical applications, the wide wavelength spectrum is acceptable and is quite useful. However, in many cases it is desirable to narrow the wavelength range of light for specific applications that require a selected region of colour or frequency. This task can easily be accomplished through the use of specialised filters that transmit some wavelengths and selectively absorb, reflect, refract, or diffract unwanted wavelengths.

Colours may be defined in objective physics, and accurately specified as specific electromagnetic frequencies in the visible-light range. The objective components of colour was first proposed in the late 1600's by the Dutch physicist, Christian Huygens, as a source of radiant energy that travels in a medium and hits an object that reflects and absorbs different portions of the light spectrum. Colours may also be described in subjective terms, as sensed by a vision system, perceived and experienced by an individual or an artificial camera device. Nowadays, it is also an easy task to process the colours and even manipulate the environment through devices, like sun glasses, filters or even by modifying colours in digital photographic pictures by computer programs.

The subjective components of colours are related to the visual perception, i.e., the cones and rods in the retina of the eye, that are connected to the brain that further interprets the information received as colours and generates sensations in response to that information. The camera device receives the energy in a similar manner, through pixels located on an optical electronic chip. The human perception carries visual information from the eye to the visual cortex of the brain, where the experience of colour is made conscious and human emotions, associations, and memory are generated. This process is related to our earlier experience and makes us in some sense also unique, since each and every individual have their own experience that influence what they perceive. In the artificial camera vision system there is a similar process of data collection, handling and analysis that has to be taken care of before presenting a picture on a screen.

Also, the human stereo color vision is a very complex process that is not completely understood, despite centuries of intense study and investigation. A complex vision process involves the paralell interaction of the two eyes and the brain through a network of neurons. The proceess of communicating colours to the brain, and the degree of subjective involvements needed when reconstructing the visual view is a complex process. The human perceptual sensing has restrictions in detecting colours and nyances, e.g., when it is getting dark, and is still not fully understood. Thus, there is no reason to doubt that there may exist other colours which humans are not able to detect and identify. However, if the human capacity is restricted and we are not able to experience visual sensation out of our visual

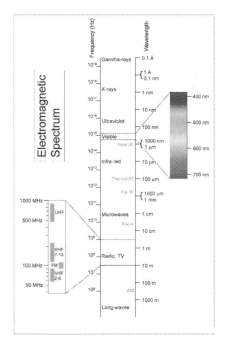

Figure 2.5. The electromagnetic spectrum also known as the visible light spectrum. Image source: http://commons.wikimedia.org/wiki/File:Electromagnetic-Spectrum.png.

range, then we can rely on new technical and emerging systems, e.g., the night vision systems built on IR-technology. Actually, important distinction of these technologies will not allow us to see new colours, instead the artificial devices allow us to get a glimpse into a different wavelength or range.

The shades of colours are, as much of the human efforts to sort and classify different states, often described in standardised terms. A colour can be described in different shades, which are related to a general classification system of colours. In the text below, some useful colour related terms are explained to show that there is an attempt to structure and organise the varieties of existing colours.

- *Hue* is the local colour of an object, defined in the seven major colours: red, orange, yellow, green, blue, indigo, and violet.
- A *value* is the degree of lightness or darkness of a hue, defined as a light related value added before a colour.
- *Saturation* is the relative intensity of a hue when compared to grey.
- *Temperature* is the term describing whether a colour is relativly warm or cool. Colours containing of blue, green, or purple or undertones thereof are cool colours. Similarly, colours containing red, orange, or yellow or undertones thereof are warm colours. The concept of color temperature is of importance in photography and digital imaging.

2.2.1 Colour Deficiency

Colour deficiency affects an individual's ability in the sense that he may identify the colour spectrum slightly different than another person. The degree of difference in experiencing the colour qualities may, of course, vary between individuals and the variety of colour sensation individually perceived can also be a limitation in daily life. But nature also plays an important role in experiencing the perception of colours, when we find huge varieties and nuances in different shades and places. The sensory effects of colours are also dependent on their context, indicating that the colours are in a way dependent on each other. This effect can be shown when experiencing that a particular red colour may appear more intensely red when relatively compared to an adjacent colour of green than if it is surrounded with grey. Also the surrounding light intensity relative to the colour and seen from different angles will affect our perception.

Most people will, when ageing, have decreased vision ability, mainly due to natural reasons, like age related factors such as deteriorated pupil size. This normally results in an effect whereby the incoming light to the retina may have reduced ability to discern contrasts, Winn (1994). Ageing also affects colour vision in the sense that the vision effect is reduced in discriminating between different blue and blue-green colours. This effect is believed to be caused when the lens turns slightly yellow, which will drastically reduce perception, as it will affect the abilities to distinguish between colours.

The human eye sensations arise when light stimulates the retina. The retina is made-up of sensing units, as mentioned earlier, called rods and cones. The rods, located in the peripheral retina, provide us with our night vision, but cannot distinguish color. Cones, located in the center of the retina, are on the other hand not effective at night, but sensitive in perceiving colours during daylight conditions.

The colour sensing cones each contain a light sensitive pigment, which is sensitive over the visual range of wavelength. Our genes contain the coding instructions for these pigments, and if wrong coding instructions are structured, then the wrong pigments will be produced, resulting in the fact that the cones will be sensitive to different wavelengths of light. This process then may result in a visual acuity. The colours that we perceive are completely dependent on the sensitivity ranges of the genetic instructions for those pigments, and if not correct then the visual sensations may result in a colour deficiency. There are many different types and degrees of colorblindness – or more correctly called colour vision deficiencies. This means that colour deficiency normally implies that some colour sensation is absent and the person in some cases may not be aware of the deficiency.

People with normal set of cones and light sensitive pigment are able to see all the different basic colours and their nuances combinations by using cones that are sensitive to one of three wavelengths of light – red, green, and blue. Figure 2.6 shows the "normal" colour scale standardisation.

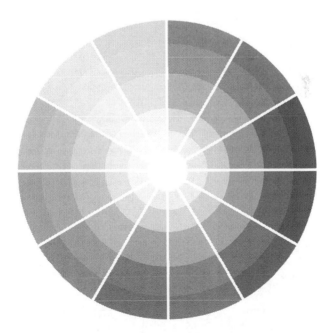

Figure 2.6. The standardised scale of a "normal" colour distribution.

2.2.2 A "Normal"Colour Scale

The following arguments are intended to raise the question of colour normality in a population, by showing a frequent variety in the human population. The colour perception may then exhibit a variation between individuals, that in normal daily life will not be discovered or create any noticeable problems. However, there should be a general understanding that a number of people do not perceive the same colour sensation, in for example, commercial and advertising feature in a tv program. The quality of colour identification can indeed be increased by categories of colour deficiency people, that could be supported by an additional artificial system. A complementary sensor system that compensates and correlates and strengthens or weakens the colour shades into colours that can be perceived by an individual's conditions and needs, is considered to very be useful.

A small colour deficiency is present when one or more of the three cones light sensitive pigments are not quite sensitive and their peak sensitivity is shifted. A major colour deficiency is present when one or more of the cones light sensitive pigments is not providing "normal" responses.

Substantial number of people in a population that exhibits a colour deficiency. It is estimated that between 5% to 8% of male humans and 0.5% of female humans are born with a colour deficiency. An excellent overview of this subject can be found in Neitz (2001).

Further, it is shown that one male out of 100 statistically has a colour deficiency (protanomaly) that is prominent under certain hard driving conditions, e.g., intense sunlight, rain, or foggy weather conditions, where there is a risk for these individuals to mistake the colours of a traffic light.

Protanomalous or deuteranomalous colour deficiency individuals can usually pass as normal participants when taking part in everyday activities. The 5% of males (deuteranomaly) may make occasional errors in colour names, or may encounter difficulties in discriminating small differences in colours, but usually they perform well. Actually these groups may not even be aware that their colour perception is in any way different from normal. However, these deficiencies may affect certain situations, for example, in harsh light conditions.

Dichromasy males normally know they have a colour vision problem. For this 2% of the population, colour deficiency may affect their daily life. They see no perceptible difference between red, orange, yellow, and green. This group exhibits a different colour experience compared to the normal viewer, when certain number of colours appear to be the same colour. The observer actually senses a uniform view and this deficiency can be effective when determining the status of the traffic lights when driving in conditions where it is hard to localise the physical position of the light.

There obviously exists an important and influencial colour deficiency in the population that simply cannot be ignored when discussing the human perceptual senses. This wide variety in colour sensations of population is indeed a factor of importance when pushing forward for a complementary sensing ability that may increase the individual's integrity and increase the awareness of identifying correct rendering colours. In a society where colours are of importance, this also seems to be a safety aspect in situations where colour perception is required.

2.2.3 *The Colour of Water*

If we look at earth from the space we explore, it is as a big water reservoir with a huge water paradise. This may be true, but if we consider the actual available water for human disposal, in areas as for example personal use, food production, etc., we have to consider the fact that 97.4% of the water is salt water. Of the existing fresh water, only 0.8% is available for human consumption. The rest of the freshwater is bound at the poles, permanent snow and in the glaciers. The fresh water will surely in the future, be an issue of importance in the world to secure the living standard as the resources have to be equally shared between the population. More than 1.2 billion people today have no access to safe drinking water. The proposed and complementary drinking water test device, described later in this book, may be one step for pushing the safety of externally deciding whether the water is drinkable or not.

Another essential fact is that water is not equally divided between the population. Nine countries have 60% of the earth's fresh water resources within their boundaries. The difference in the fresh water allocation is not only unevenly distributed, but also unequally perceived by individuals.

Water basically has a light blue colour, which however may shift into a colour spectrum, from deeper blue as the depth of the observed water increases to green when affected by other components in the water, e.g., minerals or algae. The blue colour is caused by selective absorption and scattering of the light spectrum, Fig. 2.7. Impurities dissolved or suspended in water may give water different coloured appearances.

Large reservoirs of water such as oceans elucidate the water's inherent slightly blue colour, not as was once believed a reflection of the blue colour from the sky. The main reason why the ocean is blue is because water itself is blue-coloured. Optical scattering from water molecules provides a second source of the blue colour, but coloured light caused by scattering only becomes significant with extremely pure water. According to the frequency spectra for pure liquid water, a short water column has an obvious light shade of turquoise blue. Thicker layers (many meters) appear as much darker blue colour or the colour of blue-green water set against the stark backdrop of Victoria Glacier, as can be seen in Fig. 2.8. It is only when collected in a large body that water's blue colour is less affected by external sources, and becomes apparent.

Scattering from suspended particles also plays an important role in the colour of lakes and oceans. A few tens of meters of water will absorb all light, so without scattering, all bodies of water would appear black. Because most lakes and oceans contain suspended living matter and mineral particles, i.e., coloured dissolved organic matter, light from above is reflected upwards. Scattering from suspended

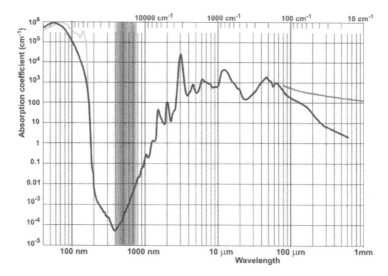

Figure 2.7. Absorption spectrum of water. Image courtesy of Martin Chaplin. © 2010 Martin Chaplin.

Figure 2.8. Lake Louise, Banff National Park, Rocky Mountains, Canada. Photo courtesy and copyright Peter Wide © 2010.

particles would normally give a white colour, as with snow, but because the light first passes through many meters of blue-coloured water, the scattered light is appears as a blue colour. In extremely pure water, as is found in some mountain lakes, where scattering from white coloured particles is absent, the scattering from water molecules themselves also contributes to the blue colour.

The colour of water in rivers, lakes and oceans is mainly affected by particles and solutes, which can absorb the light. The specific colour of water can be observed, when the consistence of water is changed to snow and ice, as an intense blue colour is scattered back, without any external interference, e.g. can be experienced from holes in fresh snow. Hues of blue to a change into a slightly green nuance are also scattered back when light is scattered from frozen water, for example from the mountainside or a glacier wall.

Mountains lakes and rivers may be of a turquoise colour at a distance, but the pure water produces a light scattering light blue colour that can be observed, when reflected at the surface. The sea and lake surface colour may also be affected by the reflected skylight and together with the depth of the lake and the angle the observation occurs, have a strong impact on the observed colour. Changes in ocean colour can be caused by a variety of sources. The indirect ocean colour may also understand and follow the effects of climate changes. For example, phytoplankton has a clear impact on the biological system and these organisms can be a large scale indicator of the present conditions, such as the carbon dioxide used in the photosynthesis process, that provides almost half the oxygen we breathe.

On the coastal water, another type of particle usually affects the light-scattering properties and the colour of the water. The amount of particles suspended in the water, the type of particles, and the depth and clarity of the water all contribute to the overall colours observed along the coastline.

2.2.4 *The Variety of Normal Colour Perception*

The beautiful pictures in the previous section are indeed a sensation for the observer. However, how do we actually communicate the colour perception and the sensation we have experienced? The term colour perception performance has obviously an important task in co-ordinating a general protocol in communicating with other humans. In a meeting between people, the interaction in the argumentation and the inspiring discussion is essential. The communication, which is often developed in certain directions with specific rules of behaviours, e.g. an illustrative body language or relating the discussions in a convincing manner. An arising question would challenge the statement if there was an obvious power of diversity effecting the discussion, on the individual basis, i.e., a statistical perceptual spreading in a population's performance. For example in case of colour diversity, there seems to be worth to rise whether a growing part, e.g. the older generation, of the population is not able to detect the right colours. For example, the colours may in certain situations not be as overwhelming as other state when admiring a beautiful colour of water in a mountain lake. The stating of "a normal visual perception" may be very restricted when we may not be able to get the full flavour of understanding the nature we are a part of. In the light of nature, there seems that the human performance is less flavoured than many of the animal's sensational behaviour. There is a possibility that humans actually do not envision the important vision sensation as for example the visual abilities of, e.g. a cat, bat or bee. There

Figure 2.9. A beautiful colour view of the Great Barrier Reef. Image courtesy of Nick Jenkins. © 2010 Nick Jenkins. All rights reserved.

is also an obvious risk that the variation in sensation performance in a population does not envision the important sensations.

It seems that our strongly limited colour perception is far from the performance of species that has evolved specific properties to further develop the physical eye vision capabilities. Perhaps the solution can be to adopt to an artificial interface, where the use of technology improvements may increase the performance and provide us with a richer and more flavoured life. However, when minimising our efforts to develop our abilities on our own by using supportive artificial abilities we do it at the sacrifice of taking an active part in human evolution.

The reality may be that the individual may not experience the colours in a masterpiece of art, like a painting with colourful expressed stroke of the brush without external supportive devices. Maybe painters have used the illusions to explore the contrast of colours by generating spatial relationships. The techniques of expressing colours are generally based on the use of colour temperature by advance warm colours that will recede the cold colours.

This technique has been used by painters, since the time of Leonardo da Vinci. He used this technique to create a special colour expression in the painting Mona Lisa. In this painting, the woman's figure is composed primarily of warm hues, while the background is composed primarily of cooler blues and greens. This distinction in colour temperature may have an intention to place the person in advance of the background, and may have the intention to explore colours in relation to each other. The intention to strengthen parts in a painting or any indeed other subjects may not be appreciated by people with certain colour deficiency.

The painters are known to use specific techniques to explore the varieties in the nature and to discover the intensity in the surrounding colours, e.g. finding the

Figure 2.10. A surrealistic colourful computer-generated using mathematical algorithms painting. Image cortesy of Fredrik Wallinder. © 2010 Fredrik Wallinder. All rights reserved.

specific light and colour circumstances when it is blowing, cloudy weather or time of the day. An old prescription technique is given in Minnaert (1954) often used by the painters to increase the colour richness of the landscape. By observing the scene with the head downwards bending forward and looking between the legs a new and intense experience will be exploring the colours. The intense feeling of colours is supposed to be related to the increased flow of blood into the head. On the other hand, a minor blood flow is reported to occur when lying on one side on the ground but the effect may still be experienced.

2.3 APPLICATIONS

It is very useful to get an initial insight about, and a basic knowledge how, the world around us is perceived and how we are conceptualised by this content. Nature may show how situational sensations actually are presented to us and refined distinctions of another are received through our human senses. It is very enlightening to have at least a minimum awareness regarding the discrepancies that may occur in the sensation of nature that will affect our perception. Physical qualities may affect the final impression differently, depending on who is observing the phenomenon. In the following section, a number of simple experiments are shown to get the right feeling of the human capability and what distortions that may occur in the perception process when observing the nature. In connection to this approach, the after-word may be useful to read and to reflect upon the contents regarding the human's correlation to nature.

2.3.1 The Human body in an Unnatural Environment

This illustration will show what discrepancies may occur with the perceptual senses when interacting with surroundings that are not our natural habitat and therefore not effectively adapted to our sensing organs. To put the human body under water can be an experience to exhibit extraordinary indistinct and a feeling of a hazy and unclear situation. The human sensing capability perceive different than its normal conditions and a feeling of unnatural perception of an unusual situation outside our normal habitat is experienced. This situation will affect our senses in a different manner. The hearing and smelling capabilities are blocked, the pressure on the skin, actually the whole body, indicates an abnormal situation. When we open our eyes, we experience that our visual system has some discrepancy, that is the picture we experienced under water is blurred and the objects seem fuzzy. Under water the light refracts differently than in the air and the eye loses its ability to focus.

When the body is in the air, it is the outer surface of the eye, the cornea, which collects the rays of light and causes the formation of an image on the retina. This is only supported slightly by the crystalline lens. However, under water the cornea is neutralised, because the fluid refractive index in the eye and the water outside it are nearly equal. This optical phenomenon causes the eye to affect the rays to go straight on without any refraction influence at the bounding surface of the cornea and we may see it as a difference from the normal mode as illustrated in Fig. 2.11.

The illustration has now demonstrated an excellent means of judging how insufficient the crystalline lens would work, if that alone was responsible for the

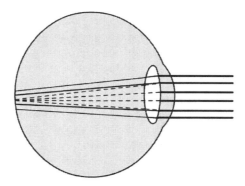

Figure 2.11. Visual impressions by formations in the eyes; (a) in the air. and (b) under the water.

formation of images. This is a far-sighted behaviour indicating that focussing under water at a point of light remains equally fuzzy at whatever distance it may be. In order, to at least identify an object under water is probably to keep it as close to the eyes, in order to get a large angle to the object. To make a comparison, a coin in clear water becomes visible at an arm's length, however, a piece of iron wire is invisible to detect under water at any distance. An additional problem is also, that there is a substantial difficulty to estimate distances under water. These experiments show that the humans are vulnerable, when entering and acting in new and unexplored environment. The perception may be if less useful in these prerequisites, as the habitat has in the long term been shaping our perception, in order to get the best possible conditions to succeed in an unsecured but natural habitat.

A personal test can, with advantages be conducted, to get the feeling of how the world is explored from under the water. This experiment can indeed verify the vision perception behaviour of the physical properties in the eye and lead to specific feeling of insufficient perceptual abilities. The light from the air has a quality, when entering the water surface, that the light is changed when the ray makes an angle smaller than 45 degrees with the vertical surface. This phenomenon will result in a large disc above the head (which can be noticed if one is under the water and looking up). If looking sideways (i.e., more than 45 degrees from vertical line) the ray will be completely reflected at the surface and a mirrored image of the ground will be seen.

Recent research indicates that the visual defect under water can be improved by training and we able to see more clearly Gislen (2003). This can be seen in ocean nomadic people in Southeast Asia who spend in an essential part of time under water. However, children with an intensive underwater training can increase their vision abilities under water, Gislen (2006). This study indicates the possibility to

Figure 2.12. The eyes of a fish. Photo courtesy and copyright Peter Wide © 2010.

Figure 2.13. Human exploration in space aiming for settlements.

enhance the sensing performance of the perceptual organs when entering new and unnatural environments.

We may just speculate that the eyes of a fish will explore a spectacular view by using the effective perception that makes it experience its specific surrounding under water, i.e., in its habitat as seen in Fig. 2.12. Further, how the world above the surface is seen from below or what the visual sensation is if the fish may appear in the air, are hypothetical questions that may never be answered.

The human visual eyes are a result of evolution, with specific preferences required within the surrounding atmosphere, which seems to be an important factor for the visual state of performance. The light scattered in an interface of air with specific light conditions, moisture and climate has evolved the human eye into a fine tuned organ that is effective in the specific environment. However, since we lately spend more time in artificial short–distance environments, this is probably changing our future evolution strategy. This direction may also be affected if future activities are explored, for example into space as seen in Fig. 2.13, where a new environment probably will interfere with the evolutionary direction and make lasting impression.

Chapter Three

The Perception

The definition of the word *perception* from the aspect of natural science may be expressed as:

physical sensation interpreted in the light of experience

(Encyclopædia Britannica - Merriam-Webster Dictionary). This definition, however, does not take into account the massive and complex structure that humans perceive with all their senses. The definition below may be considered more general and involves the smelling and tasting abilities, which in no sense can be considered as physical sensations, since they are based on chemical substances.

The interpretation of sensory information using both the raw data detected by the senses and previous experiences

(Oxford Reference Online - A Dictionary of Biology in Biological Science). However, one of the most describing definitions of the human perception process can be a slightly modified version of the definition found in Encarta Encyclopedia.

The description of the word perception is:

perceiving is the process of using the senses to acquire information about the surrounding environment or situation ; within the range of human perception.

The psychological related definition of the human perception may be related to a neurological process of observation and interpretation through the sensing organs. Any neurological process of acquiring and mentally interpreting information from the human senses is concerning the;

- recognition and interpretation of sensory stimuli, based mainly on capacity of the memory.
- insight, intuition, or knowledge gained by perceiving.
- the capacity for such insight regarding the handling of a huge information flow.

3.1 INTRODUCTION

In psychology, and also cognitive sciences, the meaning of the perception concept, is the process of attaining awareness or the understanding of sensory information

Figure 3.1. The perception visualised in a picture showing the beauty of nature. Image source http://commons.wikimedia.org/wiki/File:Bierstadt_Lake,_Rocky_Mountain_National_Park,_USA.jpg.

as defined earlier. However, it is a complex process, not yet fully understood. The word *perception* comes from the Latin *perception, percepio*, and is related to the understanding of the processes concerning receiving, collecting, and taking active part, in understanding the sensation concept. The sensation is apprehended with the mind or senses and how the human, interprets with these sensations. Perceptual sensations are experiences that consequently receive, collect and further take an active part in processing the sensation by the brain.

In the technology approach of a corresponding artificial sensory system, we define a human-based sensing capability as a device in close interaction with the individual. The main concept is to improve or complement the human sensory perception with additional sensor information. This process will then, most likely, provide a richer information in real-time and a more adequate perception that make more accurate decisions and conclusions. This will then provide an increased sensational performance. A critical process is, however, how the information is perceived to the human in order to be considered as a communication related to the interactive perception process. Then, the information is considered to be of a process in nature, e.g., an open and effective communication between the human perceptual sensing and the artificial complementary sensor system.

The technology development in this field of human related perceptual devices has for the last decades often been focussed on mimicking a single perceptual system, e.g., the vision system. The goal has often been to perceive similar solutions of mimicking a single human sensation. Also auditory perceptual sensation devices have been developed by supportive hearing aid, e.g., the cochlea implants. However, in the last few years, a number of additional technical perceptual solutions have been presented in the literature concerning touch sensors and odour detection. However, all these solutions exhibit a single perceptual device and in comparison with the human capability, are considered as being on a very rudimentary level. These sensations of independent technical approaches are doubtfully able to perform as advanced multimodal devices, interacting with humans and at the right time providing information in tightened and effective standards.

The far most important and remaining task of research activities to be performed is to combine the symbiotic effects between human perception and an additional sensor system providing a true multimodal sensing capability. Further, a vital crucial aspect is the connection from the externally collected information, to provide the communication in an attractive strategy that is not occupying the human perceptual resources, i.e., to be adaptable to the interaction of the brain.

The complexity of an artificial and complementary sensory system may be unattainable, by understanding more about how people with disabilities experience the world around them. For example, what is the sensation of a deaf-blind person and with what premises does the person experience the world. We know that other perceptual senses are directed to be more sensitive and fine-tuned with regard to the ability to compensate for the arising sensory loss. But we hardy know how or even if there is compensation made to that part of the population exhibiting a lack of minor deficiency. Does a sensing acuity perceive, even, if not optimally may be partly compensated by another sensing ability, when exploring the complex worldview? Further, following this line of arguments, do we have a variety of perceptual apprehensions in the population, where for example, colours are not a standardised measure, but more estimated by the individual's subjective relation? This question is essential since it may also vary between modes and times of a perception.

Following this line, are we able to create a complex, but still a creative mixture of structural procedure? When also including earlier experiences that influence the perceptive "picture" of experiencing the beauty or essence of the world, we actually are able to sense. To continue with this aspect,

- can an engineering student in the last year of his study get interested in understanding the subject in its entirety, with that specific and maybe unique individual subjective and personality experience attained earlier in life?
- can this student meet with the views of a technical and abstract application in appreciating the intricacies of engineering as an art or a logical function, i.e., by selecting only those aspects that are relevant and labelled with qualitative and in some aspect lasting impressions?
- can a former student who is now a retired expert understand all the gained experience attained earlier in life as a unique piece of subjective and personality treasure that will now slowly fade away?

One of the most impressive perceptual processes may be the sensation of writing words into a context that may be different for each reader. The success of reaching the reader's mind and to bring the reader into a technical description, a mathematical equation or an abstract reasoning mode for an abstract view of mediating the perception of the author's sensation. This process is also affected by the reader's ability to recall the feeling that occurs in the text. The text is correlated to earlier experiences, or if not trying to adjust their subjective level of earlier understanding of the subject, e.g., the feeling of solving an abstract, and complex mathematical equation.

The integration between the two adherent activities, but still separable processes, sensing and processing activities in the human perception is complex and probably individually adapted. The detailed procedure of how we, in detail, perceive the surroundings with the senses and based from the acquired information actually take an active part in a perceptual sensation procedure, is still not clear. The interplay between observation to perceive dynamic information, perceptual action when actively collecting data and reaction based on the acquired information, makes us unique and unpredictable as individual subjects. As stated earlier

Figure 3.2. Complex logistics create a need for structural and effective storage. Photo courtesy and copyright Peter Wide © 2010.

in this section, perception is a complex process, involving both sensing organs and the brain that is subjective and indeed a procedure not fully understood. This fact may affect an individual's personal behaviour and at the first glance verify that an aged person with a huge amount of earlier experience at disposal would make more efficient decisions than of a younger person without this storage of knowledge. The fact however, indicates that memory is divided into a short and a long term memory, that may have a fuzzy priority of what is important knowledge and for how long to keep it in memory is important, Best (1995) and Eysenck (1993) as illustrated in Fig. 3.2. The performance of the sensing organ also decreases over time and the limiting age-related observation and action abilities may in some situations be compensated with a larger amount of experience at least until the dementia effect starts to interfere.

3.2 PERCEPTUAL MODES

These illustrations show the enormous power of the memory capacity, which momentarily insights into how an individual's perceptions, feelings and thoughts could become a reality of a person's decisions, actions and general behaviour. A major research area for the next century for science, is probably to discover how experiences can be translated into processes of understanding observed sensations, apprehended with the mind process and how the mind interprets these sensations. These perceptions are frequently used in different situations. We have to comprise the perceptual behaviour, in deciding and follow what the given behaviour psychology or social rules and codes is consistent in order to act and interact according to the surrounding code.

3.2.1 Social Perception

Social perception is considered to be the perceiving of attributes, characteristics, and behaviour of a social environment, Augoustinos (2006). The quality of social perception is the process of forming impressions of an individual or a group of people as exemplified in Fig. 3.3. The formed perception is based on available information of the environment, previous attitudes about the perceptual stimuli, mixed with a current portion of stress, sensitivity and the mood. When forming impression of other individuals or groups, we have a tendency to work under certain social perception biases. For example, in the social bias, named the Halo effect, humans tend to perceive a nice person as being good in the sense that the person possess desirable personality features such as friendliness, sociability and intelligence than person(s) considered as less attractive. A positive perceptual effect gives rise to subsequently positive qualities. Indeed the expression "You never get a second chance to make the first impression" in this context is quite true.

In another type of social bias, named the Rosenthal effect, we tend to make distinctions about the strength of the influence of our expectations. If we consider that our bias about other people is confirmed, we then tend to operate under certain bias perception.

Figure 3.3. Social pattern is a natural process in society, however, it is influenced to a certain degree of social bias. Photo courtesy and copyright Jesper Johansson © 2010.

Our expectations control what we pay attention to, and what seems to be for the sake of convenience, the choice is not to perceive information that is not in coherence with our expectations and earlier experience.

When growing up, we quickly learn to adopt specific socially accepted behavioural rules and codes. From the media and during school lessons, we learn the social perception patterns we need to absorb, to conveniently sort people, groups and populations into categories. Later in life, as adults, we may understand that these generalisations of detailed simplification have a social implication that is maintained from generation to generation by the interpretation of a social perception.

In social perception, different factors determine what we pay attention to. The external affecting factors that influence the resolved and directed attention, may be gender, clothes, appearance and age. In addition to the internal affecting factors, e.g., gestures, voice characteristics, body carriage and facial expression, result in how and why we notice a certain person. Also, a portion of our own sphere of interest, expectation, as well as other influences contribute to the overall impression.

The social perception is indeed a general element that is present in our daily life. The perception is often biased by other parameters than the sensations received by our senses. The social perception is in general also affected by our expectations, may be depending on external conditions and impressions experienced earlier. The importance of a social structure, for example the organisation at work, normally gains importance by focussing on how people in a working organisation feel and are satisfied compared to not being tangible and being able to receive social rewards, such as for people being unemployed.

3.2.1.1 Culinary Arts

The social perception in the perspective of a dinner, meal or other types of food consumption is a social concept relating to the environment and other individuals, showing that some sort of relationship is established. Food alone does not create the entirety of a meal. Every meal has a message and communicates a certain feeling to those who take part in it. Thus, the substantial entirety of a meal is like art, seeking to inspire and to overshadow any feeling of "commodity" that it might have, Watz (2008). Of course there are other possible perspectives regarding how, when and where we consume food, but the meal may well be used as an experience to increase the sensation, and by that means also increase the quality of life.

In Edwards (2008), a proposed methodology for a dinner is presented. The dimension of the dinner consists of a complete, integrated experience composed of five separate aspects that has to fully integrate in order to achieve the complete and lasting feeling of an individual sensation. This form of social perception is comprised in the academic field, of culinary art. The five parts which compose the full sense of entirety is according to the following proposed methodology:

- the room,
- the meeting,
- the product,
- the management control system, and the
- atmosphere.

The full flavour of the peculiar feeling that encompasses a successful dinner at a restaurant or in a special family dinner is believed to be a joint feeling, where all the above aspects of entirety is individually perceived. Also contributing effects are the joint feeling of well-being between participating individuals, e.g., friends around the table or in the room. The social and health aspect of a successful activity is of high emphasis and is a ground layer for fruitful meetings between individuals common experiences.

The general trends in society tend to create influences that affect the perception of what we consume in different social situations as shown in Fig. 3.4. The food industry is continuously adapting to the trends, as well as the social requirements in a society and tries to correspond to these demands. In the field of food science, food analysis is an important area to ensure the adaptability of the product to the

Figure 3.4. The Nobel dinner is a distinguished part of the Nobel ceremony. Image courtesy of Fabian Seitz. © 2010 Fabian Seitz. All rights reserved.

customers needs and requirements. The food industry uses food analysis as a powerful tool for the development of new or improved products and the exploration of consumer preferences and acceptances within the field of marketing, Lawless (1999). A commonly used definition of sensory analysis is referred to as a scientific method used to evoke, measure, analyse, and interpret those responses to products as perceived through the senses of sight, smell, touch, taste and hearing. The sensory analysis concerns the human sensory system related expectations of a specific product, and may be seen as a perceptual verification of the expected sensations. This can be performed by human panels comprising both experts and consumers, as well as by sensory analysis instruments (or a combination of these). The food sensory sciences, on the other hand, involve some kind of a three parts relationship between a specific product, the consuming person involved and a social relationship to the environment, including other individuals sharing this perception. This relationship could be seen as a joint three-interface process. For example, the synergies focus on the relationship between the chemical properties of products in a product or meal and their sensory sensations. Important factors also concern regarding how an individual consumer perceives the product, and finally how the social perception is experienced by interacting with the environment, Fig. 3.5. A meal experience is also dependent on the atmosphere in the room, verifying that other people have the same feeling and hopefully enjoy the situation, like friends around a dinner table. The five characteristics of the meal can be seen as a model to define the entirety of the experience by contributing to define the social perception when enjoying the company of family or friends.

The three interfaces described in a sensory analysis are a very useful combina-

Figure 3.5. A cafe may be the perfect location where the atmosphere creates a balance between a meeting and food. Photo courtesy of Shawn Lipowski. © 2010 Shawn Lipowski. All rights reserved.

tion of a successful social interaction. An essential part of the sensory analysis is the description of the sensory characteristics of a product, which could be applied within the restaurant sector, as well as in food stores. To present a descriptive expression labelled for describing the specific flavour, for example, a fruit by describing the characteristics of an apple maybe useful. The verbal expression is an important feature, when communicating experiences between individuals. It is often easier to describe sensing features, like the appearance, sound and texture of a product than the individual and subjective value of flavours. The best argument for selling wine at a restaurant is still to let a sommelier directly influence the customer to experience the wine by smelling, tasting and perceiving the flavour, Manske (2005).

The expected effects of a developed sensory language are essential in describing the flavour from fruits and vegetables. Describing the characteristics of fresh products is a neglected service to the customer today. For example, the taste and aroma descriptions for wine, extend to comprise the relationship to the consumption of food, and based on the expectations of consumer's attitude, preference and consumption will probably increase the understanding and interest of the food

Figure 3.6. The description of the taste of an apple should be presented as a customer service. Photo courtesy and copyright Peter Wide © 2010.

consumed. If this emerging tendency of increased awareness of food qualities continues, then the validity of the proposed methodology will indeed have a positive effect on both the companies' productivity, as well as positive effect on the health of the population.

3.2.2 Psychology Perception

Perception is one of the oldest fields in psychology. The Weber-Fechner law in psychology originated in the middle of the 19th century. Also the recent and more flexible Stevens law, from the 1950's, has been frequently used. Both laws are mainly founded on experiments, where two stimuli are given and a person is tested to find out whether the intensity between the stimuli is noticed. The theories, which quantify the relationship between the intensity of physical stimuli and their perceptual effects, for example, in perception of temperature, moisture and comfort in clothing, Li, (2005). The human perception is often related in the literature as perception, when comprising both social and medical/psychological aspects. The research models are often related back to Gestalt theory, Palmer (1990), and they still have an extensive impact on the thinking around human perception.

In order to receive information from the environment we are equipped with sense organs, e.g., eyes, ears and a nose. Each sense organ is part of the sensory perception system, which receives sensory inputs and transmits sensory information to the brain. A particular problem for psychologists is to explain the process, by which the physical energy received by sense organs forms the basis of perceptual experience. Sensory inputs organisation are obviously converted into perceptions of desks and computers, flowers and buildings, cars and planes; into sights, sounds, smells, taste and touch experiences. The field of visual perception is further presented in Gibson (1987).

Psychologists distinguish between two types of processes in perception: bottom-up and top-down processing. Bottom-up processing, is also known as data-driven processing, because perception begins with the stimulus itself. On the other hand, top-down processing refers to the use of contextual information in pattern recognition, that is the beliefs, cognition and expectation in perception. The top-down processing is also known as a conceptually-driven process, Eysenck (1995) and emphasises importance of the information provided by the visual environment in a perception process. These two lines in visual processing, comprises the fact that huge amount of information is collected by the eye, but most of the data disappears by the time it reaches consciousness in the brain. Norretranders (1998), estimates that about 90% of information is considered as not prioritised data and therefore not considered in the continuous process. Therefore, the brain has to conclude a person's momentary visual perceptual impression input, and added on past experience and knowledge in the sense that a person's actively build his/her perception of reality. The individual's perceptions of the world are obviously real time sensing inputs, that are aggregated with subjective hypotheses based on past experiences and stored information. Sensory receptors receive information from the environment, which is then transported and further combined, with previously stored information about the world, which the person has built up as a result of earlier experiences. The formation of incorrect hypotheses, as well as incorrect sensory perception will lead to distortion and provide a degree of acuity of perceiving the world, e.g. the visual sensation.

A major theoretical issue on which psychologists are divided is the extent to which perception relies directly on the information present in the stimulus. Some argue that perceptual processes are not direct, but depend on the perceiver's expectations and previous knowledge, as well as the information available in the stimulus itself. This controversy is discussed with respect to the degree of involvement of the perceiver's earlier experiences and previous knowledge.

The expected conclusion so far would seem to confirm that indeed we do interpret the information that we receive, in other words, perception seems to exhibit features of both data and conceptual drive processes. However, the influence and balance of involvement by internal brain correction may be related to the individual's momentary situation, e.g., similar experience, mood, stress, etc. This mode of sensing ability may also vary in a manner that is not even foreseen, when creating the discriminating human being as illustrated in Fig. 3.6.

3.2.2.1 Human Emotion

Following the statements of Damasio, (1999), then the expression;

feelings

in its definite form is reserved to our private emotions or mental experience of a perception. In practical terms, this means that no one can observe or identify the feeling of another person but on the other hand we can be wide open and anyone can observe our feelings when we perceive the state of our emotional feelings. Through the emotions we can express ourselves, both consciously and unconsciously, our thought and feelings that are reflected into the physical world. Emotion is a complex process, which is based on different psychological function levels. The term feeling designates to the subjective experience of the emotion, e.g., joy, desire, love, security, sadness, fear, etc. It has sometimes been described that feelings have a lust and dislike characteristic and can provide an additional value to our experiences. Human feeling enables properties such as comfort and wellness, and should truly be measured in a population. A sincere involvement of human emotions in a population will probably increase communication activities resulting in an active sense of emotions, e.g., body language. This may result in an improved quality of life.

In daily life, the emotional effects can never be underestimated as a promoting competition effect. As performance quality, technical specifications and the price of products are often similar, nowadays, the emotional qualities of products become more and more important. The measurement of human emotions enables the designer to produce devices with a highly effective interface between humans and machines (artificial system) (HMI).

The quantification of emotions are of interest to enable a sort of "emotional standard" that can be a common platform. However, before one can measure this state, the emotions have to be properly identified and characterised. When it comes to defining emotions, there are a number of constructive ideas and theories. Researchers from different scientific fields are interested in human emotional behaviours. Different disciplines of scientists, such as physiologists, neurologist, biologist, biochemist, etc., have speculated on the evolution, structure and character of emotions. Psychologists, anthropologists and sociologists have proliferation theories about emotions and their significance to the individual and society. Philosophers are deeply concerned about the specific and detailed understanding of the views on emotions. Creative artists have proposed explanations for emotions, their meaning and impact, and ways to portray them in sculptures and paintings. Other disciplines also have their views on emotions, including political sciences, economics, performing arts, etc.

The very beautiful, satisfactory and emotional phrase,

I feel good

can be considered, according to Damasio, (1999), as a background feeling that arises from the corresponding background emotion. That is, background emotions are not any of the so-called universal emotions of fear, anger, sadness, disgust, surprise and happiness, neither the so-called secondary nor social emotions. Background emotions are, according to Damasio, (1994), sometimes of low-grade, although sometimes quite intense emotions that are not in the forefront of the mind. Feelings originating from these emotions are considered to be of importance, since they flavour a person's live by the way they define the mental state and colour.

The field of emotions is experienced as an emerging interest in science, that are concerned with the importance by which emotions have an evident effect on the human mental processes, that also deeply affects other related technological and scientific disciplines. One of the areas affected by this emotional interest is the scientific approach to the user-centred design of products and environments. During the past years, there has been a significant paradigm shift in the role of human factors in the study of human and product interaction. Increased recognition is now being given to "emotional-based" human factors looking at expectations, dreams, values and aspirations, and emphasizing the holistic nature of the target interaction, e.g., the individual, the product or the environment. Many techniques have been developed to increase the performance that can be considered under the product development phase or educational courses, like emotional engineering or effective design. Specific programs are being directed to increase the knowledge about emotional engineering, i.e., capturing and focussing on the subjective and emotional needs of consumers and incorporating them into the design process. An essential and may be the most successful methodologies in this field of knowledge is provided in Nagamachi, (1995), that originated in Japan in the early 70's and aims to translate consumers' perception expressed in words into design elements. This methodology allows design and evaluation of products before launching them into the market. Product Semantics, presented in Osgood, (1957), with the descriptive title "the measurement of meaning" is still after many decades representing effective and powerful techniques to measure the emotional aspects of a product. Application examples is shown in Matsubara (1997), Jindo (1997), and Alcántara (2005a and 2005b). Other measurement methods based on pictures have also been shown in Desmet, (2003).

The area of emotional engineering is an important factor in the development regularity of designs. The area focusses on the evaluation of engineered products and does actually make a person understand the connections from emotions as a result of the human perception and of course by mixing values of what a person has earlier experienced. Many recent products reflect the view of buyers preferences and an increasingly important factor is also to identify the personality who has made the decision to make the purchase, e.g., the importance of the interior design in a car when including the built in coffee-mug holder. The reality seems to be that people in some countries frequently spend time in the car consuming coffee and expect, or at least appreciate, a cup holder.

3.2.2.2 Human Attention

In psychology and the cognitive sciences, a vital process of human perception is the power of attaining awareness, that arise when an individual is actively taking part in the available sensory information. There is a probable statement, that the focus on awareness has underestimated the richness of sensory information available to perceive in the real world. The reason can be related to the huge information flow that is constantly perceived through our sensory organs and the very small fraction, about 1%, that a person is aware of, Norretranders (1998). The human perception also seems to exhibit a huge information filter that with preset preferences, or earlier experiences and knowledge, and provides a person with a refined focus of situational awareness.

Just as one object or situation can give rise to multiple perceptive responses, an object may also fail to give rise to any perceptual responses at all. That is, if the sensation has no grounding in a person's earlier experience, or of any other interest, then the person may literally not perceive it. This process is called the passive perception, Goldstein (2006).

The crucial term in our daily life is attention that according to Endsley, (2000), is an essential process of situational awareness. Despite many definitions, the notions to express knowledge that there are different ways of processing sensory organs input in the brain, accentuating some aspects of the surrounding scene and ignoring others. As the visual organ may provide as much as 80–85% of the sensory inputs, we usually use phrases when saying "to see is to attend", Yantis (2003). For instance, due to the crucial involvement of attention in visual awareness, and "change blindness" effects of "inattentional blindness" can prevent objects from being seen despite being directly fixated on, Rensink 2004), Velichkovsky (2002).

There are two basic theories of perception: passive and active perception. The passive perception process can be compared to a person who is observing the environment without taking any action. The person, is collecting information mainly by the vision senses and passively observing the scene. In this mode, the person does not take any specific action in mind or other expectations connected to the input sensations. The individual is simply acting as an observer, passively observing the envionment, Biel (2002). The opposite process is of course the active perception process, where the individual is actively taking part in the situations in the environment and provides decision and actions thereof. The active person is taking part in the happenings in the environment and providing a situational awareness that is probably making the person an important player in his surroundings as illustrated in Fig. 3.7.

By adding the active perception principle to an artificial system, the "intelligence" in the system provides a possibility to reformulate a mission and change the behaviour, or system parameters, according to dynamic situations in the environment. This property facilitates the possibility to redirect the sensation process to the area of interest, Freeman (1999). The active perception paradigm was orginaly intended to be applied in the vision systems when introduced by Bajcsy, (1985).

3.2.2.3 Body Language

It is widely accepted that body language is a powerful psychological source of information reflecting emotions and intentions in our daily life when interacting with people. An individual's body language may influence us within a part of a second, but during that time we have already formed an opinion about the

Figure 3.7. The interaction of situational awareness between the zoo-keeper and the crocodile. Photo courtesy and copyright Peter Wide © 2010.

person, Van den Stock (2007). The response may be a feeling of welcome, suspicion, attraction or disgust depending on the response received by the encountered body perception. The topic on body recognition indicates that that there is an overwhelming direction towards face recognition in the literature. An emerging interest is to explain a person's stationary mode and in some sense also the person's intent by identifying their behaviour. For example, security personnel are trained to identify bodily behaviours and follow a set of cues that are inconsistent. The abnormal behaviour of a person acting strange by showing a nervous, calm or moving in a certain pattern structure may create an increased and suspicious interest from artificial, intelligent and distributed surveillance systems, Valera (2005). These examples certainly illustrate the circumstances that by applying our intuitive understanding, as a measurement for a distinct identification of body language, will simply not be enough. To establish a person's characteristic body language, we need to explore more dimensions in order to identify a person's behaviour, by identifying an abnormal body language.

The technical advances in camera supervision of travellers in airports, the individual behaviour pattern in a crowd or in specific areas has been of emerging interest and systems have been developed, which have also been reported in the literature, e.g., Valera (2005).

3.2.2.4 Lack of Perceptual Senses

In the daily life, we exhibit an extensive but still highly normal interplay with the world around us, gather information about our surroundings and interact with it through our actions. The interaction between perception and action is complex and of high performance. On the other hand, the perceptual information is the basis for decision making and therefore sensitive for a proper reactive action ability. Perceptual deficits are often devastating in the overall performance and may lead to the obvious deterioration in decisions and response action, which in the long run also have an influence on the survival impact. An example concerning manipulation of the haptic perception in virtual environments is presented in De La Torre (2008).

We recall that the human sensing is among the last functions that abandon us when there is a life-threatening situation and danger to the body. Even when we are unconscious, the sensing abilities can be functioning. This quality may be a heritage from the past, where the body focusses on vital organs when the body is life threatened in a critical situation. Therefore the allot of sensing abilities, regarded as vital function of the body's survival nature is convincing where the perception abilities seems to be of high priority.

Medical tests have showed that if we close the perceptual sensation and prevent most of the sensory inputs entering the organs, then we are lacking the essential information to be able to act normally in the surrounding and in actual, we cannot handle the normal situation in life. The person, after lacking essential sensing input information, begins to fill in the information gaps by hallucinating. This situation to deliberately get rid of a person's sensing abilities by outmanoeuvre perceptual organs has frequently been used, e.g., in prisons. The individual is deprived of his contact with the environment and the world reference frame gets fuzzy. This mode creates internal thoughts about "where am I, what is around me" and when it is constantly dark, the individual may loose time reference. Then, if the aim is to increase the confusion by adding external sensory information, e.g., sound, smell or light disturbances in close contact with the person, in order to create illusions that can be related to earlier experiences or unpleasant situations. The breaking down of a person's perceptual intimate, which may also be its fundamental relation to the surroundings, is considered of high internal security, that easily can be affected by manipulation in forms of depriving a person by artificially affecting their sensation and by tampering with their perception, Reynolds (2008). The result might be devastating for the individual by affecting the interference of the perceptual structure that may create long-term effects on the human mental health.

3.2.2.5 Increased Sensory Perception

The human sensory abilities are highly dynamic and complex. Perception is highly effective and normally not performed in a static mode. The normal perception when exploring our environment is that we move our sensory organs, head and body at different distances in such a way that the sensory information is adjusted to the present situation. The perception then has to follow a co-ordinated pattern, e.g., looking at the person who talks, getting closer when listening to a person in a buzzing environment, etc. The complex co-ordinating structure is redundant in the sense that the complex information received from the sensory organs always occur in the same way, when we act and move our body in the environment.

We are normally equipped with a sensitive and non-distorted sensing ability from birth that provides us with an increased sensory perception when it is non-distorted at its best performance. This specific ability is fragile, and has to be provided with a process to initiate the sensing structure. This activity begins with the infant's view of its surroundings, in order to learn and build-up a knowledge base of experience. Studies show that new-born infants are capable of showing shape constancy, Slater (1985), recognise individual faces, Turati (2006), prefer their mother's voice to other voices, DeCasper (1980), appreciate skin-to-skin contact that has a positive caring effect, Erlandsson (2007), as well as the auditory, tactile, visual and vestibular intervention have a progression on alertness and feeding in pre-term infants, White-Traut (2002). This suggests that perception can be well established from the beginning and provide surrounding information and perceptions already from birth.

The complex ensemble of exceptional impressions that we experience in our daily life normally provides us with a fruitful interaction with pleasure and meaning to enrich our existence. However, also in this case the variation may affect the personality of an individual. Some individuals may exhibit an extra sensitivity to sensory inputs that experience an increased sensory capability. Sometimes this perception is undesired sensations that will give rise to negative impressions, i.e., the individual feels a discomfort and illness. This behaviour can be seen in people affected by perfumes or are hypersensitive to radiation from electrical wires and objects. However, we may also find this category of increased sensitive individuals making an advantage of this quality by working in environments like whiskey distilleries, the perfume industry or simply in the industrial food sector. The extra sensitivity makes it unique in establishing the quality and provides the fine-tuning of the sensing qualities in products that are specially appreciated by people with ordinary sensing capabilities.

Thus, this group of hypersensitive individuals often have explicit problems in communicating the extra sensitivity feeling of subjective impression to other persons and in many cases this will cause an effect of being discriminated in social life.

3.3 SUMMARY

The perceived impressions of an environment, and the possible effects which bias the sensory input, give rise to certain questions which are related to how to proceed in the concept of artificial human sensory system. The approach concerned is to facilitate a symbiotic design and create a structure that complements the ordinary sensing capabilities. The most urgent design however, is concerning one part of a population to recover their missing or reduced sensing capacity. A sensor system that complements missing or reduced sensing information is vital for this category in order to provide impaired sensing abilities. This approach of concept also involves the aim of a vital social influence on increasing perception and the possible action in certain situations.

A possible approach for blind individuals to use the body skin as the interface when receiving information from an artificial build-in spectacles, processed and further interfaced with the user. The area of a room can, as illustrated, be framed on the body skin. Obstacle, and the person's location in the room can be communicated in the tactile sensory. The examples illustrated a possible interaction strategy between a supportive system and an individual person in a human-system interaction as illustrated in Fig. 3.8.

This approach will probably give rise to a number of questions that each of them needs to adequately answer. The following questions can be asked:

On the other hand, a key question is whether we, as humans, exhibit an increased true capability when supporting our perceptual information with complementary artificial sensor data?

Do we manage to handle the external information and integrate well-defined specific and pertinent data on the right spot time?

Will actually the individual make more precise, suitable actions and well-founded decisions or will it lead to information overload in the long run and create a more vulnerable individual? Do we consider it a pleasure in receiving additional and artificial based information?

And finally, do we get an enriched life by supporting the human-based sensing ability?

In other words, are the proposed artificial based sensing process able to, result in abilities that increase the enriched human sensing capabilities and, would it be able to provide an individual with not only an extensive world contact, but also support to prolong the human's quality of life.

In a common situation, where people for example, need additional and artificial sensations to compensate for seasonal affective disorder that can be illustrated in activities of light therapy for people suffering from harsh winters and lack of sunlight during part of the year, Rosenthal (1998). The effect of light therapy has in recent years been questioned as the evidence, is regarded insufficient to determine the real effect of light therapy, SBU report (2007). Another argument for increased perceptual added sensations are stated when people experience music for enhanced health, mental stimulation or simply relaxation, for example the famous "Mozart effect" in Rauscher (1993), but are also criticised in Jenkins (2001). Nelson (2008) states that music may play an important role as an adjunct therapy in critical care. Also reported, an increased relaxation was established in medical operations by both doctors and patients, by reducing, e.g., blood pressure, heart rate, stress hormones and amount of pain medicine to be taken, Conrad (2008), or enhancing learning, Jausovec (2006). Both examples show, even if further studies are necessary, that induced additive perception are seen as illustrative processes for increasing the perceptual sensations that provide an additive feeling in specific situations. If these actions taken will provide elderly people to get the possibility

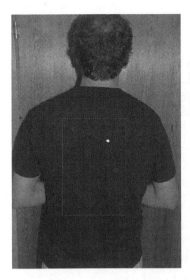

Figure 3.8. The possible interaction strategy between a supportive system and an individual person in a human-system interaction. Photo courtesy and copyright Peter Wide © 2010.

to complement, their normally-reduced sensing abilities often caused by age, with the ability to stay active a longer period in life. By that means there is a fair chance that an active perception driven life style will influence a positive dislocation in time of age related disorders. If artificial sensors in the future will be able to provide people by giving them the current sensing impressions and the right stimulus, then the sensor systems are indeed demanding a correct complement to the lack of natural-based sensations.

It is believed that by affecting our perception with an enriched amount of stimuli, we will be able to increase our perceived input. Even if there exists some doubt of the obtained effect, in combination with limited sufficient medical evidence and confirmation of the physical increased value, there is surely evidence of the effect in the traditional methods of treatment that is pursued in different parts of the world, Jouper (2008). The evidenced-based effects are often reported in convincing results, even if there are uncertainties sometimes in the results presented, Jenkins (2001). The fact is however, that when enriching the stimuli of our sensing capabilities a natural consequence would be an achieved medical effect, e.g., the tactile sensing by massage. The stimuli of an enriched sensing ability is able to provide a positive effect on the body resulting in a healthy healing effect. The opposite is, the negative effect achieving a destructive effect, as described in the earlier section, when people were lacking sensory stimuli. The importance of sensory perception has obviously a varying approach in different cultures, as well as among different groups in a population. Since many generations, a continuous fine-tuning of the experience and knowledge behind the perceptual treatment process has been going on, in order to achieve an improved mental health and well-being in our daily lives, Jouper (2006). We have indeed a common view of the individual approach of perception when sensing the world, but at the end, there seems to be a strong connection to our cultural behaviours and traditional experiences. The appreciation of and valuation in the stimuli given to us will too often be neglected because the lifestyle of today is, with some exception, a life in decadent with respect to perception performance. We are today simply not using our full sensing capacity when appearing in situations that for some generations ago would have been dangerous, if not obtaining the ability to perceive the right information in time.

3.4 APPLICATION

The character of the human perception has been demonstrated earlier in the book, mainly with the objective to bring about the fact that a variation of the perceptual capacity exists and possible consequences thereof. This is then, of course further biased by the internal mixing of an interaction with impressions and situations experienced earlier. This perception process seems to be of an individual-based characteristic. The perceptual performance seems to be multidimensional in capability and the subjective ability is inherent as a unique variation for every person. This phenomenon provides the multidimensional aspects in a population indicating that a person is unique and no one is similar in the perception, decision-making and the action taken in different situations.

It may therefore be of interest to know the specific behaviour of oneself in different situations. Maybe there is a possibility to understand the underlying tendencies of how and why *an individual* perceives in a specific situation. A common situation with a daily frequency of occurrence can be the interaction during the intake of food. Therefore, the situational assessment process, when having a meal, can be of importance and affecting the general mood, e.g., the degree of pleasure when consuming the meal. The gathering together with a tasty choice of food, will in different meal situations, most likely give rise to sensational preferences. As examples, these situations provide different perception effects and are listed below:

- consuming a meal by yourself in your kitchen.
- eating dinner with your family.
- having lunch in fast food restaurant.
- eating an effective lunch in the car, or other transportation systems, in order to save time.
- having a meal outside in the park, at the forest or at the seaside alone or with friends.
- enjoying a cosy restaurant where you get your favourite cuisine, with the companionship of your best friend(s).

The pleasure of a specific meal can be decided directly after finishing the meal and without any logical reasoning and explanations you may sense the indicative feeling and maybe make a conclusion. The summary of the experienced occasion may provide each participant with a feeling of pleasure that will be kept in memory. The occasion will, if successful, also facilitate the possibility of creating a new gathering in similar situations.

The awareness of how a person comes up with that specific conclusion in a common and daily situation may be an experience of interest. Situations that a person may cope with, and maybe can inductively conclude with some reasonable explanation, is that together with existing background knowledge provide reasons for this given specific opinion, e.g., the dinner was extremely boring. This experience may provide us with a new and understandable knowledge of why the daily actions taken, and why it is provided in a specific order.

I may, in my own world of experienced situations and prioritised sensing occasions from past, emphasise the flavours that my brain calls special attention to. For example, when I smell the flavour of newly baked buns, I experience the sensation and enjoyment from my childhood. Also, I have when travelling sometimes aversion against train transportation, due to, at least my belief, that I as a child had a feeling of unbalance showing tendencies of sickness. This was at older ages compensated by the knowledge that I probably had this perception related to earlier experiences from my childhood.

The reader may also recall the strong effects from earlier gained experiences.

What sensations do make strong recall and will provide amplification of the perceptual inputs?

- *on holidays,*
- *in meal situations,*
- *family meetings.*

This self-insight of context is of importance and the fact that earlier memorised situations may affect the final feeling also in a distorted way of trying to find similar events that may help defining an object or explaining a situation.

Chapter Four

Sensor Technologies

Sensors perform a significant part of an electronic device by providing the interface to the surrounding range of interest. The sensors may be of wireless or data-bus connection, and can consist of a huge number of miniaturised micro-technology-based solutions or state-of-the-art design that intend to deliver data in effective, consistent or intelligent applications. Most applications of today still use single sensor (or mostly a very few) in applications intended to measure quantifiable values in a restricted partially defined range of the environment. This structure may instead result in a physical property were the sensor will contribute to the specific part of the object and change its quantitative parameter(s) caused by the measurement system. For example, measuring the temperature in a sealed cup of water will probably affect the measurement if a sensor is entered into the water, due to the difference in temperatures between the sensor unit and the measured object, i.e., the water.

Artificial sensor structures, or simply sensors, perform a vital function in measurement and control applications of today, namely to provide an interface to the real world. However, the glimpse from a sensor measurement may often, only reflect the local spot of its range that it is focussed on. The sensor performance usually exhibits an artificial reflection, e.g., often in non-real-time with adherent discrepancies that makes the sensor data just artificial mimicking units of the real world, and maybe with an annoying time lag. Nevertheless, it is of importance to have enough knowledge of the sensor behaviours before considering the measurement specification – what do we want to achieve?

This chapter is intended to be an initial insight to sensor technologies before moving on to the artificial perceptual sensor and its applications. There will be an overview of the classical sensor operational principles that extensively offer an introduction to the importance of artificial sensors as an information source.

A sensing unit can, in general terms, be defined as a;

"device that is receptive to external stimuli and transforms it to communicable data ".

This definition is very broad and covers everything from human senses, for example the human eye, to gas sensors used in a fire detection system and sensors used to track the speed of a car.

Artificial Human Sensors — Science and Applications by P. Wide
Copyright © 2012 by Pan Stanford Publishing Pte Ltd
www.panstanford.com
978-981-4241-58-8

However, to further restrict this definition to the principle of a sensor device, one definition can be found in Britannia, Merriam-Webster Dictionary on the word "sensor";

"a device that responds to a physical stimulus (as heat, light, sound, pressure, magnetism, or a particular motion) and transmits a resulting impulse (as for measurement or operating a control)".

This means that a sensor device can be described as an artificial device with physical properties. The sensor measures a specific local range of measurement interest, with indirect or direct operational principle and provides the foremost task of collecting specific information from that specific part of environment.

However, to be correct in including all types of sensor principles, the refined version of the above definition may be as follows;

"a device that responds to a physical, biological or chemical stimulus and transforms it to communicable data ".

In the following sections, the imperfection, behaviours and limitations of sensors will be discussed, that make further advances of the operational principle interesting. The processing of sensor data is well-established in order to increase the performance of a sensor system. Also the benefit of combining sensors with different and complementary specifications will, in many cases increase the total information.

But as in all considerations the use of a sensor, or several sensors in a system, will be of importance, when the output performance decides the inquired level of conceivable information. As always, a detailed knowledge about performance and sensor limitations are required and can be found in for example, Wilson (2005) and Fraden (1993).

4.1 INTRODUCTION

A sensor is a device that transforms a physical, chemical, or biological stimuli into a corresponding detectable value, normally an electrical signal. The sensor output can, in principle, respond to anything that reacts in a certain manner on the stimuli. Figure 4.1 shows the transformation principal system. From an environment, a specific measurement volume influences the sensor, i.e., the sensing volume, is measured. The range of the sensor performance normally corresponds to the object or measurement parameter(s) to be measured, e.g., a volume or area. The sensor reacts on the specific environmental parameter(s), the response is detected and further processed or communicated, which is illustrated in the principles and figures below. The sensing unit can be placed directly at the sensing spot or as illustrated indirectly below from a distance.

An artificial sensor is a device that in most cases measures the stimuli indirectly, which means that the stimuli is often transformed by a physical, chemical or biological input stimuli into an indirect value corresponding to the input

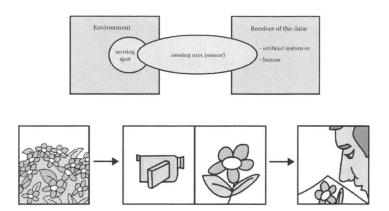

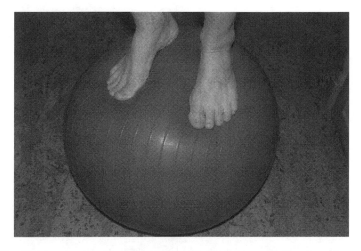

Figure 4.1. The operational principle of a sensor comprises the measurement procedure, (a), and an example of corresponding measurement organisation (b).

parameter, as shown in Fig. 4.2. The sensing part is normally integrated with an electronic signal processing unit that handles the measured environmental data by processing it further and making it suitable for the subsequent operations and interfacing. In the figure, a possible scale design principle is able to measure the person's weight, by indirect measure of the pressure in the pilates ball. The scale will then indicate whether a person has lost or gained by the indirect measurement of the pressure changes in the inside of the ball.

Sensors are an essential part of a control system. Without sensors, most technology applications would not exist. Sensors perform a basic function, namely they provide specific information from the real world. The importance of sensors

Figure 4.2. The sensor operational principle illustrated as an indirect pressure measurement of a person's weight. Photo courtesy and copyright Peter Wide © 2010.

as information collectors, however, has been compromised in the last decades due to limited development activities in sensor research compared to other electronic areas. The recent revolutionary progress of smart sensors, wireless sensors, and micro/nano-technologies, Meijer (2008), is in Huijsing (2008) regarded as the third revolution named "sensorisation", after the earlier technology revolutions "mechanisation" and "informatisation". The focus will indeed increase the emerging demands for powerful performance in intelligent systems, where the sensors are considered to be a vital part. But it has to be emphasised that the sensors themselves need to be incorporated or even integrated in active systems that effectively handle huge and extensive information in real time. A single sensor, even with whatever smartness incorporated, will not meet the demands of performance with dynamic and actively searching information abilities. However, by vastly integrating a huge number of small, intelligent and specific sensor units that are also cheap to manufacture, the performance will, of course, increase the range of useful information. A carefully designed multi-sensor system will provide the end user with more sophisticated and valued data, compared to traditional single or few sensor(s) systems.

In the present trend, comprising the emerging developments in the field of artificial human-based sensing, the main purpose is to achieve a performance in the sensor system that corresponds to the efficient, naturally complex and valued interaction between a sensor system and the human. Information has to reflect the real world status in the close human proximity, and presented to humans at the required time and in an acceptable quality. The information flow can, in this context, be seen in a similar way as a modern manufacturing plant, where products have to be delivered at a competitive cost, at the required time and in an acceptable quality to the customer. An information system also has to deliver at a competitive cost (flow), giving the cost for specific information that are used in the human decision-making process. The information has to be presented and correlated with the ordinary human's own perception obtained information in time, and before the decision is made. This process result can be seen in the view of the consequence that the sensor information really is an additional and weighted input to an efficient decision-making or to provide further active information search. Finally the desirable information would have such quality that the information will be accurate and not provide misleading information to humans or autonomous systems.

4.2 FUNDAMENTALS OF SENSORS

A sensor system is considered to be a device that acts as an interface between the physical world and a modelled world. As such, a sensor converts a physical, chemical or biological phenomenon into an indicative property, i.e., a value entering a transfer function between the real world and a corresponding artificial system model. This "value" corresponds to a real world phenomenon and usually measures the requested information in a time period or sequences and correlates to one or many reference values.

The sensor values in a system model are often the base of information that can, complemented by other sensor information measure the environment or even actively affect the close proximity by actuators. The sense-control model is a common process in the scientific field of measurement and control. According to the illustration in Fig. 4.3, the human skin is the main sensor to detect temperature, and the brain is the decision-making system that decides whether the temperature is too cold or too warm to have a shower, or if it is the right temperature. The desired temperature is of course a subjective matter to be decided by the person who takes the shower. Also, it can be noticed that a fuzzy temperature scale is normally used when a person is acting as a sensor in a control loop.

Even if we get the most perfect sensor with the best possible performance, we should bear in mind that it still could give totally erroneous information when not used properly. The sensor itself usually measures only the extracted parameter in contact with the sensing element. The parameter performance procedure can be exemplified by the fever indicator sensor that measures a temperature using infrared light in the ear of a child. More correctly, the temperature sensor only measures the temperature in the ear and maybe only in the ear at the moment it has been in contact with a warm and cosy pillow. The influence of the environment has to be considered in a measurement process. The environmental or real world effects have a strong influence on the actual information from the sensor and should be added to the overall information before a decision is made. There-

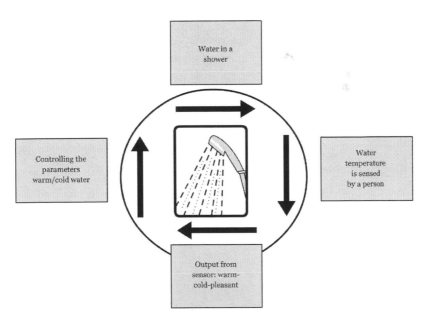

Figure 4.3. The human in a closed control loop where the human body is acting as a sensor device.

fore these affecting constraints should be considered in the sensor system model design. Thus, an attached model is of essential importance that such a model is able to collect the information we actually are interested in without unnecessary disturbances. Then, with some expected logical certainty we can be sure that the sensor system performance behaves in an expected manner.

The sensor characteristics included by the manufacturers are often limited in performance and have to be put in the specific circumstances and applications given. The sensor specification viewed in a product data sheet often exhibits external limitations, which restrict the working abilities of the sensor in the real world. Usually a single sensor has drift, aging and temperature characteristics that make it necessary to calibrate the sensor system in a structural process. At least we have to consider the possibility of adjusting the sensor result to the expected external and internal conditions.

Also, the sensor has to be designed to work in coherence with following designs and the provided output data has to be coordinated with an interface and following units, also when they are all integrated in one single electronic chip.

When considering the possible limitations in sensor performance, we should have in mind that the main purpose of sensors is to make quantifiable measurements in order to produce data that is related to a reference or different references. This statement considers the fact that by quantifying the sensor values in relation to calibrated reference values, we actually are talking about adding the basic measurements with a correlation factor for each restricted type of measurement process.

To summarise, there needs to be a consensus in the abilities and limitations regarding sensor advantages, and also the restrictions to provide, when involving them into complex design applications. Sensors perform a vital function, namely probe occurrences, and act as an interface to the environment. As also mentioned in the afterword of this book, the environment can be highly dynamic, complex and the human intelligence is not adequate in following the increasing demands and rules. Therefore, the sensor fundamental principle is an important tool in industry, research, and academia to optimally calculate the sensor performance and overall system design in coherence with the expected environmental prerequisites. The reference guide on selecting, specifying, and using the sensor in an effective design manner is of importance, in order to know how to measure external properties of measurable parameters.

4.3 TERMINOLOGY

In reality, we may generally consider the fact that all sensors have dynamical characteristics, that will create changing performance and most likely establish a sense of complication. All sensors also exhibit shortcomings, rendering errors and uncertainties in the provided output data. The imperfection of dynamical effects can be related to physical theories and sensing principles, while others are related to design, production and measuring conditions. Therefore is sound knowledge about sensors and sensor behaviours of utmost importance in designing a measurement system and predict its operational principles? This knowledge will hopefully provide the designer with the best possible performance of the sensor system, in achieving the expected information without future draw back.

To be aware of the basic terminology, and make use of the characteristics, when dealing with sensor specification and general system performance is of importance to succeed. The most important sensor characteristics are summarised below.

Range

Every sensor is designed to work over a specified range. The design range is usually fixed, and if exceeded, may result in permanent damage or destruction of a sensor. Further, by exceeding the range does not guarantee accuracy of the results without a new calibration, and this is a more likely scenario than damaging the sensor. It is customary to use sensing elements within only the part of their range, where they provide predictable performance and specific enhanced linearity.

Zero

When taking a measurement, it is necessary to start at a known value, and it is often convenient to adjust the output of the measurement indicator to zero at the reference value. It is, therefore, often a value ascribed to some previously defined point in the measurable defined range.

Zero Drift

The level of the value may vary in time from its previous set zero, when the sensor works properly. This introduces an error into the measurement equal to the amount of variation, or drift as it is usually termed. Zero drift may result from changing factors such as temperatures, electronics stabilizing, aging of the sensor or due to influence from electronic components.

Sensitivity

The sensitivity of a sensor is defined as the relationship between parameter functions of the input and output properties. Thus, sensitivity can be described as the change in output of the sensor per unit change in the parameter being measured, i.e., the minimum possible change of value. The factor may be constant over the range of the sensor (linear), or it may vary (nonlinear). The sensitivity is an important parameter to estimate the fine-tuned or sensitive value properties.

Resolution

Resolution is defined as the smallest change that can be detected by a sensor device. This fluctuation is a measure of the sensor resolution and is given in the specification as the minimum change that is detectable in the range.

Response

The time taken by a sensor to approach its true output when subjected to step input is sometimes referred to as its response time. It is more usual, however, to quote a sensor as having a flat response between specified limits of frequency. This is known as the frequency response, and it indicates if the sensor is subjected to steps, for example, similar to a sinus-oscillating input of constant amplitude, the output will faithfully reproduce a signal proportional to the input. The response function can be further related to the system bandwidth and transfer function.

Linearity

The most convenient sensor to prefer and choose in a system would of course be the one with an acceptable linear transfer function. The ideal function provides an output that is directly proportional to input over its entire range, so that the slope of a graph of output versus input describes a straight line. However, the deviation, is the actual transfer function compared to the ideal function defines the sensor linearity (or more correct the non-linearity).

Hysteresis

Hysteresis is the behaviour and sensor inability to return to the same output value when applying the same input parameter. Hysterisis refers to the characteristic that a sensor has, in being unable to repeat faithfully a measurement cycle, in the opposite direction of operation, compared to the data that have been recorded in one direction, as seen in Fig. 4.4.

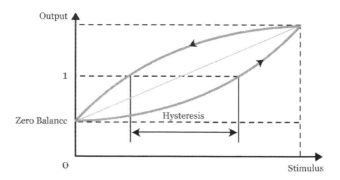

Figure 4.4. The hysteresis function of a measurement cycle.

Calibration

If a meaningful measurement is to be made, it is necessary to measure the output of the sensor device in response to an accurately known input. This correlation process is known as calibration, and the sensor devices that produce the calibration input are described as calibration standards. Calibration is a standard procedure to achieve a common reference.

Span (input)

The dynamic range or span of input properties that may be converted by a sensor is called a span or an input full scale (FS). The value FS represents the highest possible input value, which can be applied to the sensor without causing an unacceptably large inaccuracy, as illustrated in Fig. 4.5.

Full Scale Output

The full-scale output is the algebraic difference between the electrical output signals measured with maximum input stimulus and the lowest input stimulus applied. This includes the deviations from the ideal transfer function, as can be seen in Fig. 4.5.

Accuracy

A very important characteristic of a sensor device is the accuracy, which really is a measurement of inaccuracy or uncertainty. Inaccuracy is measured as a ratio of the highest deviation of a value represented by the sensor to the ideal value. It may be represented in terms of measured value of accuracy of delta, as shown in Fig. 4.5.

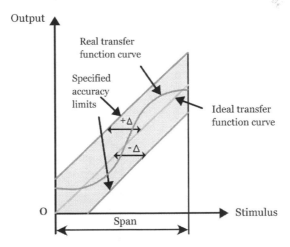

Figure 4.5. The dynamic range of a sensor. (Caption in figure to be deleted).

Sensors in the Laboratory

The previous section focussed on the internal measurement effects that may interfere with the sensor performance. However, the external interference with the measurements normally creates an effect which has to be taken into consideration. The measurement procedure is always restricted to minimise the ambient effects that influence the sensor measurements. In a specific measurement laboratory, the atmosphere is considered to be clean. Ambient characteristics provide relative elimination of external disturbances. The air is controlled and relatively free from contaminants, the temperature is kept in a stable range, the area is free from vibration, and the personnel are specially trained in the handling of measurement principles and the equipment they use. The industrial measure is often made under reverse conditions. When testing a sensor system, there is a possibility to substitute the sensor unit with a corresponding test-generating unit, i.e., a calibrating unit. The signal produced must be capable of using the ordinary system from transmission to the collecting and processing equipment included, which may be a considerable distance away. The substitution of the sensor device intends to compare the performance of the sensor devices by then using the same remaining system. The tests will have the aim to verify the performance from the view of the sensor and secure the elimination of the system errors.

4.4   BASIC SENSOR PRINCIPLES

The sensor principles are of most importance when investigating the achieved performance that is required, in order to deliver the proper capability to an electronic system. There is, in many aspects, a question of chosen sensor operational principles to appropriate applications and their demands for specificity. The system characteristics required will in most cases regulate the selection of sensing unit. The system specification including the strength to process measured information conditions, will decide the requirement level of confidence and success. The absolute minimum requirement is then to know the measurement conditions. The system requirements are dependent on several conditions, where a few of the more important sensor systems characteristics are presented below. Essential discrimination that can be related to system performance is, as given, the concepts of identification of human involvements, gathering of sensor information, indirect/direct operational principles and single/multi-sensor systems. Therefore due to its complexity, a structure of application related sensor principle toward the direction of this book, intelligent human sensors are needed.

4.4.1   Relation to Human Abilities

The classification presented here is maybe comprehensive in a multi-dimensional perspective, and challenges the traditional measurement approach. However, the proposed classification suggests a related aspect of measurement and is directed towards the main aspects of sensor categorisation. The sensor principles based on the human ability to recognise structures and patterns, and even to be compatible with human perception are indeed important factors in sensing features in complex environments. A sensor's abilities to be able to interface with the human capability that can perceive or relate a measured parameter is of vital interest, and are of the main aspects in the concept of this book, namely the artificial human sensor approach.

However, there are a number of sensing principles, where human performance is not able to cope with artificial sensors. The human perception has obvious limitations and may not sense specific parameters, or even not be able to perform a real understanding about an ongoing measurement sequence. The word "super-sensor" has been used for sensor types that have an unique potential of collecting information that humans normally cannot explore. These super-sensors are categorized as devices, whose potential in cooperation with humans, are such that their technology empower us with an extended reality, Siegel (2003).

Examples can be found in various sensing areas related to the basic human sensing systems as listed below.

— *auditory*, ultrasound sensors that perceive frequencies outside the human audible range.
— *vision*, thermal cameras that can measure invisible infrared radiation, which humans cannot perceive.
— *olfaction*, electronic noses, can measure chemical compounds that are not detectable by humans, e.g., the presence of carbon monoxide.
— *gustatory*, electronic tongues, that can taste chemical compounds in very small ranges, e.g., the presence of chlorine in drinking water.
— *kinetic*, tactile robot hands, that provide more complex gripping modes than humans.

These examples indicate that we are able to invent, design and make advanced technology related systems that are extensively able to complement the human perceptual performance. There is, however, a challenge in using these types of "super-sensors" performance as an effective and attractive equipment, with usability toward the existing human perception, in an active and interactive information process. To get individuals to fully understand these sensing abilities, there will most likely be a need for an adequate and human friendly solution. The solution has to cope with an effective interaction and translate the measured parameters into a communicative flow of redundant information to be presented to humans or human related systems. This is not a trivial problem when the information often has to correlate with basic parameter information. For example, when merging quantitative data from a temperature sensor (e.g., 38.3°C) and human fuzzy descriptions from an electronic nose device (e.g., smell of vanilla) into a reliable and fused information, in order to get a human related qualitative value (e.g., warm vanilla yogurt). Moreover, the measurement procedure has to depend on a reproducible, significant and calibrated tradition. The established qualitative value should also dependent on an individual's subjective characteristic qualities, behaviours and habit.

4.4.2 Sensing and Gathering Information

Sensing and gathering information from the world around us is regarded as a major step when collecting and organising data and a fundamental cornerstone in building the communication and interaction of intelligent systems, normally referred to as the intelligent human computer interaction. The expectation increases rapidly when using the aim of the phrase – intelligence behaviour (or intelligent systems). The often huge amount of gathered information that is needed requires a strategy, that in some adequate manner does not make use of any intelligent behaviour, when obtaining the required amount of data by fusing single sensing modalities into a multiple fusion information model. The desired depth of information is gained from the use of different physical, chemical and biological principles, that cover different information spaces to generate time related data that may have individual specificity, i.e., range, accuracy, resolution and reliability. Therefore, the strategy to properly fuse the sensed information, often from various sources and complex environmental properties, is focussed on the concept to produce the required, specific and indeed the expected amount of information in the available time. When dealing with a human related sensor system, the ultimate challenge is to design an active system to collect the most effective information and to provide an interactive communication process in a reasonable time scale.

Techniques for fusion of multi-sensor data are drawn from a diverse set of more traditional disciplines including digital signal processing, statistical estimation, control theory, artificial intelligence, and classical numerical methods, Llinas (1997). The characteristics of the mentioned, commonly used techniques have to be merged with the specified requirements needed to find a generalised sensor fusion solution for context-related applications. More details, and further reading can be found in Wu (2003).

4.4.2.1 Indirect Sensor Principle

Within the field of traditional sensor technologies, there are several different operational principles of sensing. In principle, the most frequently used technology is a common sensor technique that exhibits an indirect operational sensing element. The sensor property may indirectly use known physical, chemical or biological properties to convert the measured parameter into a more applicable and often used physical quantity before further data processing occurs, as illustrated in Fig. 4.2.

An excellent illustration of an indirect measurement is the well-known detection principle based on piezoelectric quartz crystal micro-balanced technology (QCM) used in biosensor applications, Rogers (1998). A resonant frequency, which is measured from a quartz crystal and changes output frequency as a linear function of the mass of a target substance deposited on the crystal's electrode surface, will change. An attractive detection of molecules with a high degree of specificity is commercially available in, for example, drugs and explosives, Biosensor Application (2009). In this exemplified application, as illustrated in Fig. 4.6(a) the sensor elements use an indirect measurement principle. The need to detect, analyse and compare airborne molecules at the picogram level is a requirement when measuring in this type of high-technology application. The concept is based upon the properties of which a well-defined principle was used to process the known sensing parameter that exhibits the actual transfer method. Commonly used biosensing technology principles are used in military and security applications.

The illustrated technologies have a documented reliability to work in specific environments and type of applications. Each of the technologies is vulnerable and

(i)   (ii)

(a) An indirect early waring system. Modern technology-based sensor illustrated by a piezoelectric quartz crystal micro-balanced technology (QCM). The defecting unit is shown in (ii). Reprinted with permission from www.biosensor.se. © 2010 Biosensor. All rights reserved.

(b) A direct early warning system. A bird in a cage. Image courtesy of John Harvey. © 2010 www.johnharveyphoto.com. All rights reserved.

Figure 4.6. An illustration of two cases of defection principles.

exhibits certain specific disadvantages and advantages. However, there is a general consensus that the system specification, with the known properties, makes the measurement procedure more or less standardised in the way that we can continuously make reliable measurements. Even if the biological sensor (e.g. the bird in a cage) has to be replaced when the measurement of contaminated gases reaches a certain level – Fig. 4.6(b).

4.4.2.2 Direct Sensor Principle

The measurement process using a direct operational sensing element is usually the most effective process when parameters such as simplicity and direct conversion are highly required. Direct sensing techniques, used as operational principles are seldom seen in traditional measurement tools, mainly due to the advanced electronic technology, which provides the manufacturers with electronic solutions when choosing highly effective conversion processes in the sensing elements. Traditionally independent techniques used as direct processing sensors can, for example, be experienced in the affecting properties that correspond to the influence on an environment, which an object perceives when placed in a flow. The same properties occur when applying a weather vane outside the house on a windy day. The vane will then provide the wind properties and we can measure the behaviour or just observe the sensor device, i.e., vane. The direct principle can be commonly used as a floating device that expresses the level of fluid in a tank or the surface properties of a lake. The connection to the device moves in a direct connection to the level as illustrated in Fig. 4.7.

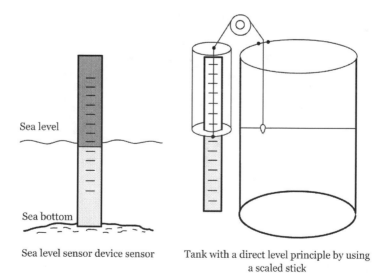

Sea level

Sea bottom

Sea level sensor device sensor

Tank with a direct level principle by using a scaled stick

Figure 4.7. Illustrating of the direct measurement principle by a floating device in a fluid tank and level of sea surface.

4.4.2.3 Single- and Multi-Sensor Systems

The use of a single sensor system is directly aiming towards the focus on the specific sensor performance and thus provides a limited understanding of the predetermined measurement goal and expected singular result. A single sensor system is probably more understandable in terms of required performance and parameters required when considering selectivity in range, sensibility, specificity or sensing features for a number of collaborative sensors achieving a common measurement goal.

A central point in a dynamic sensor system can, with advantage, be defined by the sensor specification or, in the case of multi-sensors, also by the overall sensor performance. Some measurement applications have a clear defined aim to perform single and dirigible sensor output, while others request loosely defined parameters. In reality, one single sensor is typically not able to measure all aspects of a complex entity without the combination of other sensors with different selectivity. Then the single sensor sensitivity is directed towards one particular range of interest. In case of multi-sensors, the aim is to coordinate the selectivity from different sensors in building-up a multi-dimensional selectivity space. An interesting benefit of using sensor arrays is the ability to also find vaguely defined parameters in applications, where the aim may be to detect or identify single qualitative parameters that are not clearly defined. In reality however, many sensor characteristics are indicatively responsive to several different defined parameters, where in these cases, they are considered to be non-selective. If a non-selective sensor is used alone, confusion may occur when indicating different measurement values from different defined units that can generate a similar response pattern. The drawback may then be that the measured value is difficult to relate to a certain sensor output and its state of a particular defined parameter. In the case of multi-sensors, e.g., a sensor array, a design that constitutes a multi-sensor system of locally gathered sensor elements that will provide a multi-dimensional output. A strategically designed sensor array may achieve higher performance, by using non-selective sensor elements that are able to strengthen the measurement capability, Pettersson (2008). The non-selective sensors can together measure a "picture" of the environment, corresponding to a similar picture of a mountain chain where each single measured sensor value can be expressed as a mountaintop or valley, i.e., the measured response plot containing each sensor sensitivity to a common defined parameter. The full response picture is described as the included sensors' collected values are inserted in a descriptive pattern of sensitivity. The mathematical tools used for pattern recognition can be put together in a measurement picture through a joint analysis of the sensor signals. This procedure is called multi-dimensional analysis.

The strength of complex sensor array applications aims at sensing complex parameters, e.g., qualitative values as bad odours, fresh taste or beautiful sight. The qualitative parameters that usually are requested in these measurements are in a manner connected to the human understanding and prerequisite. Normally, a single sensor may not perceive all the components of a qualitative impression. The design of a well-adjusted multi sensor-system, e.g., a multidimensional selectivity-based sensor array in close symbiosis with a straightforward pattern recognition system, is a very powerful tool. The complex system is then able not only, to provide a complete picture that corresponds to the environment of interest, but also make necessary corrections, due to error or bias in the system operational principle.

However, when using multi-sensors, two illustrative application principles can be viewed. The choice of principles is of basic interest when designing the system, and is described as follows:

— When using a number of sensors in a system with similar operational principles, we are looking at an elaborate redundant sensor system that will secure the measured feature we intend to quantify. This may be proceeded by using the same sensor in parallel, to ensure that even if one or more sensors are erroneous, then we have enough sensor power to provide the right performance. Also, we may complement the power of a sensor system with different sensor principles by using different sensor types with a variety of measurement techniques in order to get a selective range of multi-sensors measuring the same parameter. These types of multi-sensor systems are often used in highly secured functions where system failure is not an option.

— In multi-sensor systems, for example, in electronic nose applications, we aim for a broad selection of classifications of detectable compounds in the measured air volume. The selectivity is of importance to detect various chemical compounds in the measured volume. Therefore, the aim is to select a wide amount of sensors that provide a selective performance and complement each other. The multi-sensor principle will get a wide selectivity and perform a structure of quantity values that give rise to a virtual qualitative range of features. The qualitative value will provide measurements that are not detectable by only one single sensor and are expected to demand a more complex data analysis. This concept also requires a strategy for accuracy and calibration standards.

The basic terminology and primarily performance for a multi-sensor system is slightly more demanding when analysing system performance. Some of the more important characteristics can be described in the following parameters:

Response time

The multi-sensor system would be able to have a relatively high response time than a single sensor system, in order to be able to secure that each sensor has enough time to collect, convert and provide data, i.e., data processing to present a qualitative or quantitative response parameter within a predetermined time frame. In the time frame, the recovery time and the initiating time to prepare for a new measurement sample is often included.

Size

The development of production techniques that always prioritise the focus of smaller sensor sizes, is of importance when the ever-upcoming trend is to increase the number of sensing elements on a predetermined area. These minimalistic devices also often include an integration of an electronic equipment unit in the sensor package. However, considerations has to be taken of how actually the sensing profile looks like in a measurement environment, i.e., the sensing space of each sensor that contribute to an estimation of the complete measuring space, as can be illustrated in a response picture.

Cost

The costs for transfer of qualitative information from the environment into a sensor package and further communicate it to a user, are crucial factors in the overall system cost. The cost for requiring specific information has then to be correlated to other comparable information or just a lack of this feature. The overall cost to achieve specific information has always to be considered and compared to other solutions.

Power consumption

In a number of applications, there are specified limitations for the use of energy consumption. Low-power multi-sensor systems consumption will contribute to a correct measurement procedure and avoid increased risk of error, e.g., those caused by a heating effect and change in temperature, that may have a negative effect and contribute to a possible additive error. The results on minimising the energy consumption will probably also have a positive effect of decreasing the energy requirements of the overall system and are competitive in many sensitive applications, for example, in relation to humans.

Robustness

The system performance of a multi-sensor system can be illustrated by its robustness that indicate the absence of unwanted mechanical, electrical or other physical disturbances. The robustness will have an effect on the measured quantitative or qualitative value, and has to be a prioritised matter to control. The robustness is also a matter of trust of performance, where the manufacturer guarantees the performance within certain conditions and prerequisites in a specification. However, multi-sensor systems are increasingly used in measurement applications outside regulated environments, which enable a higher risk of errors.

Data processing

Even if the main request, in most cases, is to establish a stable performance, will the issue be focussed on each vulnerable single sensor element and its principles? There is also an influenced quality demand in the data itself and the processing of multiple collected quantities. The procedure of measuring a single sensor quantity, that is merged together with other sensor quantitative data into an overall response picture, imply that the total amount of data can be processed by sophisticated algorithms. The response picture comprising all the single data in a measurement cycle can further be processed in order to establish an overall qualitative correlated value. The output parameter represents the participating number of verified sensor, whose data is fused to valuable information.

Noise

Additive noise from different sources is mainly associated with randomly appearing disturbances and errors. When the sensing elements are collecting the data in their specific environment, we may presume that the collected data quantity is already contaminated with noise. The probability may be considered high, that additional noise is caused by the processing part of a multi-sensor system. The primary interest, however, is to explore, understand and find countermeasures for these peculiarities, in order to learn the pattern of the incurring noise. The aim is typically to find proper techniques to ensure that the noise will cause a minimum effect and error contribution of a measurement sequence.

Drift

The drift is defined as a gradual change of the defined sensor's output of its quantitative value under obviously constant conditions. Due to drift the outcome value will be unpredictable and systematically changed. If a number of sensors are involved in a measurement system, there maybe a need to make a change in the overall direction of an expected drift. There are some techniques to compensate for the systematic drift in a multi-sensor system. The cause of drift characteristics are complex and may be an effect from many internal and external sources, e.g., changes in the electronic surface or external effects from sunlight, temperature or change in the air pressure. Techniques are available to suppress the drift characteristics, mainly by software adjustment.

Reproducibility

In multi-sensor systems, there is a process to define the overall qualitative common value based on the contributing sensors quantitative value, when expressing the status of the momentarily features in the measurement surrounding. The sensing principles of a number of sensing elements are often selectively sensitive to each other, which become challenging when estimating the reproducibility of the overall qualitative value in the complete range of operation. In the data analysis process, a model, or calibration curve can be established, describing the relationship between the environment conditions and the qualitative response measured for the acquired quality. The calibration has to be secured in a complete multidimensional space of the system's operational range.

Reliability

The sensors are foreseen to behave according to their specifications and be predictive particularly over a long period of time. The single sensor relationship to each other is also of importance, when the selectivity and internal relations are considered in calculating the qualitative parameter of a multi-sensor system. Therefore, specific considerations have to be taken when changing one or more sensors in a multi-sensor system or replacing a complete sensor array.

Accuracy

Sensors have to exhibit a high sensitivity in a specific domain and preferably some response sensitivity for related domains. The system accuracy is created as an additive effect that depends on the different accuracies of each part of the system, e.g., the sensor elements or electronic processing unit. The drift of each sensor unit can strengthen or reduce the overall accuracy, if for example, the sensor's drift results in a positive or negative offset effect. The system accuracy has to be continuously estimated in coherence with other affecting parameters such as noise, drift, etc.

The measurement procedure of a multi-sensor system is straightforward in a sequence. When each sensor element has measured a quantitative value, the data is collected and calculation takes place. In most multi-sensor applications, the data is multi-dimensional, with the objectives of processing the data into a more presentable nature. The multi-dimensional data is often a calculated dynamic qualitative value representing the measured picture of the environment of interest, e.g., a correlation value to the human estimation. These qualitative values often perform an illustrative connection, comprised in a "picture" that may appeal as a situational awareness, such as a glimpse of the measured environment, as illustrated in Fig. 4.8. This spot of information may provide a human understanding and perception related to complement the information. This subject will be further illustrated in subsequent chapters of this book.

4.4.3 Complex Sensor Data Fusion Principle

The traditional view of a multi-sensor system may include separate sensors that are located at a distance from each other. The measured picture of the environment of interest then may be of importance when fusing data from a wide, partly measured and not aggregated environment. The distributed organisation of a sensor system network is then an excellent tool, to explain the non-homogeneous

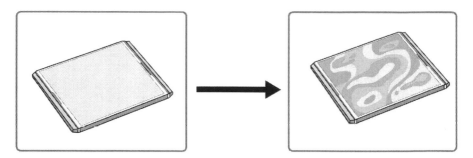

Figure 4.8. A sensor (n by n) array data response depicted as a coloured sensor response level of information picture.

information flow connected to the individually placed sensors, that provide separate data flow at different time sequences to clarify measurement inconsistency. A large number of significant applications depend on how and where the sensor elements are placed, that interface with the real world, e.g., Petersson (2008). These distributed sensor networks is an emerging topic gaining increased interest and applications may include military, medical, manufacturing, transportation, safety, and environmental planning systems.

A number of multi-sensor systems have made emphasis to be fully realised, but not completely succeeded. This is because of problems involved with collecting the right data from the specific sensors in a predetermind measuring space directly into automated systems, thus providing the requested and often time-dependend information. Sensor fusion algorithms have been a useful technique of choice for resolving these problems and to fully make use of the collected data.

4.4.3.1 Multi-sensor data fusion

The terminology in the process of association, correlation and combination of information seems confusing, Wu (2000). It seems that the phrase "sensor fusion" has shifted to "data fusion" and the trend is now moving towards "information fusion". Sensors usually provide location and time dependent flow of data that have to be further processed and interpreted in the context to provide meaningful information. A multi-sensor system may perform a data fusion process that carries out the combining and integrating activity in order to extract an added value from a diversity of measurement data, Hall (1992), Klein (1999).

A multi-sensor data fusion system has the main task to:

— organise data collection and signal processing of a number of sensors that may be of similar or different types,
— present the local and global representations, by using the received and earlier gained information,
— integrate the information from the different sensors into a dynamic and time related real time model,

— avoid wasting effort to collect and process irrelevant information,
— and possibly include some intelligent features on a higher level, regarding, e.g., the estimates, significance and interface.

A sensor data fusion system may also integrate redundant sensor data as early as possible that is in the systems lower levels. Complementary information is preferably fused at higher levels of the multi-sensor fusion architecture. The multi-sensor strategies are important parameters, concerning the overall information performance. The selection of single sensor elements in a multi-sensor architecture consider the placement in such a way as to get optimum performance during specific measurement tasks, and for the time selection of measurement activities to minimize the dynamically observed system response. The multi-sensor fusion system is typically based on an "intelligent" structure that may change strategy under processing and use the best available mode of operation at the right time. The complex system has an application independent abstraction that models the sensor function as a software description, performing its input/output characteristics including measuring process and the ability to selectively direct the measurement parameters of task specific interest.

A significant summary of the sensor fusion concept is stated in Biel (2002), that the strategy of handling sensor responses in a fusion process is comprised as follows:

> *"sensor fusion is the process of combining data in such a way that the result provides more information compared to handling each source separately".*

An extensively used method of data fusion technique particularly in human related sensing, is the multivariate analysis that comprises one phenomenon by observing or sensing a number of observations. This method is suitable when measuring more than one variable simultaneously. Further reading is recommended in Naes (1996) and Esbensen (2000).

4.4.4 The Human Role in the Sensing System

In principle, the basic technologies are comprised of sensing methodology, that often tries to mimic the human's perception of physical, chemical or biological input sensations and to convert the measured parameter. This process may be perceived in a structure, where the human perception can be involved in the measuring processing, for example as a sensing unit, processing the measured quantity value in the system, or be an active part of finding the right decision from several information sources.

In order to actively live in complex circumstances, humans have to understand the environment they are taking an active part in. Also, we have to conceptualise notions that will enable us to evaluate the impact of different natural phenomena, as well as of the impact in the interaction process of other human actions. The human in the system is considered to be an expression, indicating that in the daily interaction with the environment and other humans, we need to be active and involved in the course of events. Then we are also able to interact with artificial systems. The awareness to the environment will force us to develop the needs to communicate with other autonomous systems, e.g., other biological creatures, as well as advanced technical sensor systems, like cars and security systems.

For example, the phrase:

I feel good

is expressing a qualitative value, as illustrated in Fig. 4.9. The qualitative paradigm is among other inputs also related to the concept of quantitative values, because I have the necessary needs of quantitative values in life, like food, lodging, and transportation possibilities. Further, in the store I can buy, with money earned in doing some work, additional quantitative values in my life, with a priori defined units (money) and combine the amount of units to achieve a qualitative value. Of course the amount of units, or amount of money, should be in coherence with, and correspond to the measured and earned salary.

These interactions are small fractions of the complex intelligent control of a *multi-modal sensory perception* concerning the state of the measured process and its environment. These intelligent model-based adaptive processes provide an *intelligent* connection of the perception to action in order to achieve specified goals in advanced changing of environmental situations. These perspectives can with advantage, be used to build artificial world models. The external models built and maintained from information gathered by a multitude of sensors, that will provide a natural and complex abstraction as a representation of the state of the environment. At the perception level, the world model is analysed to infer relationships between different objects and humans, in order to assess the consequences of the proposed action, Petriu (2008).

The perception capability can be seen as a human-oriented *active window* pro-

Figure 4.9. An illustration of the statement, "I feel good and secure."

viding global information. This aspect will reduce the uncertainties about the physical world-state comprising the artificial based system, the involved human and its environment. The only *a priori* defined constraint in this model procedure is what sensing parameters are to be measured without necessarily having the knowledge about where, when or even if, they will occur. The challenge regarding this approach is concerning how multi-sensor data are integrated with human perception information, that are able to effectively be integrated in artificial world models. This approach will add a new dimension to the process, that we will denote the *perception based sensors or artificial human sensors*.

The increased interest on human-like and intelligence based system, based on powerful computer platforms, that allow for complex data processing algorithms, can be incorporated on a wider scale in new measurement methods. Virtual model based solutions are evolving even more realistically in describing the environment by providing an interactive ability in various applications, Ahmed (2007), Valdés (2003).

The model procedure enables an extensive development of a symbiotic concept involving both human and artificial sensor partnership based systems. This approach may include the use of human capability as an *explicit multi-sensor system*, as well as the use of human perception. This will act as an implicit behavioural and contextual measuring unit for indicating behavioural qualities in the environment, e.g., situation assessments.

4.5 SIGNAL PROCESSING

The signal processing mechanisms are generally performed in an electronic processor unit or external computer. The main task for a human based sensor system is to transform the measured sensing parameters into a representative qualitative value, describing the measuring object of interest. The procedure forms the basis of subsequent decisions, from sensor signals to an information status.

More or less advanced calculations are used and different techniques are utilised for the realisation of signal processing. Normally, the reason to conduct such a process includes the improvement of the interpretability regarding the environment of interest. Sensor devices in general have shortcomings that will cause imperfections of the rendered data. Under normal circumstances, many of these imperfections can be suppressed by using appropriate signal processing techniques. By incorporating statistical operations, and by analysing signals over time, it becomes possible to get proper indications when significant deviations from normal conditions occur. In the case of using several sensors, a combined analysis of multi-sensor signals may reveal hidden patterns that can be extracted and correlated to significant properties of measured samples.

Advanced signal processing is a useful tool for enhancing the information that can be comprised in the sensor data. The obtained information can enhance the usability and performance of the overall system, in order to achieve an effective handling of the signal processing.

4.5.1 Multivariate Data Analysis

Nature is multivariate, i.e., a single phenomenon normally depends on several other occurrences, and the measurement object of interest in a sensor application is expected to behave in a similar manner. Univariate methods, where a serial approach deals with one variable at a time, are rarely found in nature. This phenomenon is often of limited interest in complex data analysis. However, in all cases of measuring procedures it is a necessity to control, or at least be aware of, all these factors, in order to understand the multivariate behaviours of the dynamical sensor systems.

Multivariate data analysis in a powerful tool to increase the knowledge about complex behaviours of the operational principle and to find properties in different measurement conditions. Further, it may also include the possibility to learn more about the behaviours and how a pattern, e.g., the measured response picture, are expected to occur in future operations. Software techniques, that use multivariate analysis, have been developed and applied in numerous applications, with more or less success. The main objective is however, as always to find, explain and learn as much as possible about the sensor specifications, limitations and behaviours including the properties of the complete multi-sensor system.

When analysing multivariate techniques for classification of data, a structural approach can be used by the following road map. The analysing parameters begin with the procedure closest to the sensor unit, and may be an indicative procedure to approach when organising the system structure as given below.

— *Pre-processing*, describes the computational techniques to structure the data in a shape that is convenient to use in the following analysis. For example, normalisation of sensor data, compensation for a reference level and a common reduction of biases can result in a better performance in later analysing stages.
— *Exploratory analysis*, is considered to be an initial estimation of the measured data and provides a direction of advice of significant variables respectively possible outliers, i.e., the discrimination between measuring variables and erroneous data, the variables and noise. Suitable techniques, like for example feature selection, feature extraction can be found in the literature, Petersson (2008).
— *Classification*, is a technique to analyse processes in order to determine the existing relationship between the multi-sensor data, i.e., a set of independent, selective single sensor variables, and their relationship to each other, in other words a set of dependent classes, Esbensen (2000). A number of classification models exist. For example, statistical techniques by Multiple Regression Technique, MLR, Partial Least Sqares, PLS, or Cluster Analysis, CA, is further described in, e.g., Batagelj (2006), Loutfi (2006). However, for non-linear

models, frequently used in artificial sensor system examples, ANN and Fuzzy Based Techniques, Loutfi (2006), have frequently been illustrated in classification applications. Also the emerging interest for using wavelet transformation, Thuillard (2001), in data analysis is gaining an increasing interest in artificial sensing applications, e.g., in food quality analysis, Robertsson (2005).

— *Validation* is the process to test if an appropriate level of complexity has been attained and the requested generalised capability of the model is sufficiently dynamic. Basically, the validation process has the intent to confirm by experiments or tests, i.e., evaluation techniques, the precise extent to which a particular analysing method has the inquired character. An important aspect is to find a model whose complexity matches the structure of the measurement task. In a basic set, validation is typically performed by building a model by using a set of training data and then estimating the behaviour of that model on a second set of data, the validation data. The validation data then consists of samples, measured independently of the training data, however, with the same equipment and external conditions, Fisher (2007).

— *System Intelligence* or maybe more correctly, a system's adaptability capabilities (self-adapting models) may be of interest to adopt the principles that have been earned by learning from a set of observations. It is not only possible to generalise and apply the gained knowledge but also to correlate, predict and autonomously correct other, previously unknown, observations. The fundamental assumption made by using such principles is that the properties of a sensor may experience changes, that are induced by long or short term drift, reproducibility or unknown external circumstances. Then, the basic importance of assuming a system's stability can be maintained by an "intelligent" adaptation of unforeseen degradation in performance, by for instance miss-interpretations of the data. The degree of autonomous behaviour that cope with uncertainty may be an expression of intelligence in a multi-sensor system capability.

The proposed road map for a multi-dimensional sensor system describes a procedure increasing the security and stability in an overall system performance. The different analysing stages, as described in the suggested structural road map, may with advantage be used as separate processes or in other combinations than those proposed. The main purpose is to always recognise the ability of the system performance, i.e., to find variables, whose trends can be explained and from that experience, gain knowledge of the system peculiarities and the dynamical effects of the external environment. By knowing the system qualities, then there will be a natural possibility to logical explanation and understanding of erroneous occurrences that may happen.

As stated earlier, this book has in no way the intention, to explain the different possible procedures when using pattern recognition, but rather direct interested readers to a number of illustrative references. General concepts related to identification and learning from data is given in Theodoridis (2006), Kroenke

(2005) and Kroenke (2009).

4.6 APPLICATIONS

The human participation in the sensor system often exhibits a complex interaction between the acting individual who more or less interferes in the communicating process with the dynamic world around. The interaction is of a complex nature and our senses are often correlated to other human expressions and expectations, indicating that in our daily interaction with the environment including other humans, we also perceive volubility in the communication of speech, body language or if we act in an unusual manner. The mode we exhibit at that specific moment of action, may create a feeling of satisfaction, security, uncertainty, or instability for some reason. For example, the complexity in a common phrase:

I feel secure,

can give rise to a psychological chain of perceptual inputs:

— when for example, the yellow indication lamp is flashing on the instrument panel.
— in the car.
— I just borrowed from a friend.
— driving my pregnant wife to the hospital.
— a stormy night.
— in an unknown environment.
— without any navigating support.
— and because I lost my spectacles.
— etc.

The hypothesis states that I am secure in the situation because I know, or at least is certain about, that the yellow light from the lamp on the instrumentation panel is not flashing red. A red light normally indicates a higher degree of alarming warning then the yellow light.

However, since I am suffering from a colour deficiency, it then seems that the lamp emits a yellow-like light, but my main concern is that it is not flashing because then it would be more a serious indication, almost an alarm.

The feeling of security makes my other senses work in a normal mode that can be perceived by my wife and the way I act when managing the car in these exceptional driving conditions. The same procedure is repeated when I arrive at the hospital and a calm and confidence-inspiring doctor tells us that there is no problem. Indeed, the human communication is complex and relates to the basic behaviours and the way I act will of course affect the communication with other receiving agents in a close proximity.

My background as en engineer and the experience that I had gained earlier from this type of problem which caused the lamp to flash, made me feel confident, even without spectacles on. However, the causes that made the indication lamp to flash could also give rise to other and more serious problems such as causing the car to stop. But since I was in a great mood the whole day, it probably contributed to a deep feeling of satisfaction and I was obviously in balance. I really felt good.

This fictatious story indicates that decisions are made, and based on, a received perceptual degree of confidence in the environment and also incorporating experiences gained earlier in time, i.e., activities experienced both in short and long time aspects. Normal perceptual deficiency that may occur is in many cases compensated by experience. The main issue is, however, despite a cogent knowledge about the limited abilities of present and available perceptual capabilities, compensation affection normally occurs by other sensations, and complementing my own arising limitations. The car manufacturer has placed the red alarming indicator lamps at a location different from the yellow warning indicator lamps. By knowing where the flashing light in the instrument panel is located, the conclusion made was that the yellow light was most likely flashing. Further, despite my loss of correctional lenses, the trip went rather safely, when the car navigation system was also equipped with an artificial voice, that after pressing on some buttons was able to communicate the driving directions in a friendly and pleasant voice.

The story has mainly one goal, to ensure each and everyone that human capacity is unique and a valuable resource to rely on when we end-up in uncertain situations. These situations are by some reason also often technology-related matters or more correctly lack of technology performance. An additional point may therefore also be, to have confidence in the human communication abilities when interacting with other people as well as artificial devices.

It is however, of importance to know your perceptual limitations in order to estimate the individual adaptation in various situations. The perceptual acuity may be compensating by additive functions that improve the overall abilities. This can be done by simple corrections, e.g., spectacles, coclea implants or sophisticated support systems, e.g., electrical wheelchair. Also the need for more intelligent multi-sensor systems that complement the human abilities may strengthen the sensing functions, e.g., in preventing elderly people from falling, stop a car when driver is drunk, when affected by toxic effects, just tired or recieving a daily personal health diagnostics. The advanced systems may increase the quality of the daily work and make better use of existing support systems.

This can be illustrated by the definition, whether a person is considering himself as not feeling well. The determination of a person in being sick, e.g., having an increased body temperature is considered within certain temperature ranges. If temperature exceeds 41°C then it may be time to cool down or if below 33°C then the person may have to be warmed-up. If the person ends-up outside the fever temperature range, a life-threatening situation occurs. Now, a "normal" temperature may vary distinctly, for example between approximately 36.5°C–37.4° C depending on influencing factors such as activity, external environment, etc.

On the other hand, depending on the location of where the body temperature is measured, the temperature will vary with the magnitude of degrees, e.g., in the mouth compared to the extremities. The different parts of the body, as in the creation of the temperature scale, has historically been of significant interest, for example Newton (in 1701) suggested a scale with two fixed points, one zero point when ice is melting and the second point at the armpit of a "healthy Englishman". Also Fahrenheit suggested a second point for his 96° scale as the temperature of a "healthy man", Fraden (1993). (However, in 1742, a Swedish astronomy professor Anders Celsius suggested a 100° scale with the fixed melting point of ice and of boiling water.)

Therefore, the indirect result of a temperature change will affect our awareness and make us feel as sick as the temperature diverges from "normal temperature". The quantitative value, i.e., the difference between actual and normal body temperature, gives rise to a qualitative value of the momentary feeling, e.g., I feel really sick.

Also in this situation of being aware of your internal ability, it is of importance to know the personal perceptual characteristics, in order to make an individual adaptation and adjust to various situations. The perceptual acuity may be compensating by additive functions that improve the single abilities which also improve the general perceptual impression.

For example, if a doctor tells you that the slightly increased temperature mainly depends on the fact that it is very hot in the bedroom (measured environment). But additional indications like red eyes, a red nose and an illustrative body language probably provide in a more indicative point that you may have caught a cold.

The cause and effects of a situation where the human is playing an active part may not always be related and the connections may not be obvious, as it seems. The main understanding in interacting with a complex system, e.g., a car, train or plane may in the long run be a question of how we manage the situation and to what degree we have control of the situation. As moving with incredible speeds, that is indeed not adapted to the human perception, onboard fast trains or airplanes then our perception tells us that we need (or at least try) to have control. Even the pilots in the cockpit do not have direct perceptional control of the complete physical properties when flying an airplane. They are extensively depending on the systems' performance of transforming the physical properties to instruments and systems' ability to communicate with the operators. But also the interaction between systems in demanding situations requires an adaptable communication protocol. In literature, human perceptual acuity has been shown to be an increasing challenge, e.g., Kikukawa (1999). This indicates the problem of human interaction with advanced sensor systems without considering the human limitations and indeed the disparity that occur in an individual's perceptual sensing.

The applications illustrated intend to provide the understanding of human limitations and behaviour in order to maintain and maybe also strengthen the individual's awareness of the context. Too many misunderstandings and mistakes occurs due to specific situations where individual performance is considered uniform, and also that we have a general understanding in the solving of specific situational situations. The consideration of simple and monotonous situational assessments is in many situations treated as boring and makes us not focussed on the situation. On the other hand, when our perceptual organs are not coping with the realistic situations, we still think we may master the situation and provide adequate solutions showing that we are in control of the situation and will solve the task. The believe that we are able to cope with situations out of our perceptual range without a realistic approach to the available performance is in many cases causing hazardous situations. That is, in general we are not able to cope with extreme situations and may not fully be aware of the situational effects. Therefore, in relating to Nietzsche, the bottom line of this chapter may be;

When being and acting as a structured creature in a chaotic world, where the expectation is focussed on behaving in a predetermined and structural order. To use human perception capabilities in symbiotic cooperation with advanced sensor systems, whether it is in processes as sensor inputs, processing or communicating results, there is an absolute condition that the human perception behaves in a certain and structural manner. Then of course, we need to know how we behave in different situational contexts.

Chapter Five

An Artificial Perceptual Model Approach

The present trends of today often demand an increasing need for an extensive and directed information flow. This flow of information includes a fundamental understanding that sensors have to produce specific data to be delivered at the required time and in an acceptable quality to the user in order to be updated and be useful. There is a probable foresighted scenario that in the future, will the amount of requested information be highly competitive, delivered on time and attain high enough quality. The sensor system will then also emphasise on the dynamical and intelligent properties that will actively search for the requested information and communicate to symbiotic units, e.g., an individual. The proper planning and effective control of a sensor system, that provides essential and requested information at the proper time of delivery, is an absolute condition to achieve a smooth interaction. A proper interaction between the user and the units that collect the required information, is a critical transference necessity to perform a real-time integration system between the information collecting sensor level and the information integrating system level in order to achieve the best possible use of information in decisions. Therefore, the need for a generic model approach is obvious.

Describing the human involvement, the term "context aware human computer interaction" has been frequently used. The context and motivation is to react appropriately to the momentary circumstances that occur constantly when used extensively and implicitly, situational or context information, in interacting with humans. The context aware human computer interaction, Smailagic (2001), Wu (2003), refers to computing modes that are able to discover and react to changes in the environment they are situated. The context aware human computer interface then may increase the understanding of the reactive environment in order to achieve a higher abstraction level, and aim for a more effective human interaction.

Adjacent to the research area of context aware human computer interaction, is the natural relation to the human awareness by expanding the model representing the awareness of the reactive world to include the interaction with an artificial sensor system. The human interfacing is a crucial step and one of the most fundamental tasks in building an effective and rational human-system interaction.

Artificial Human Sensors — Science and Applications by P. Wide
Copyright © 2012 by Pan Stanford Publishing Pte Ltd
www.panstanford.com
978-981-4241-58-8

The context aware behavioral system has to make use of a new paradigm. The approach has, to relate to an interfacing sensor system that provides symbiotic information at a level of human capability and artificial intelligence. Since the individual usually finds her own information, which is based on the person's gained perceptual ability, with the highest priority, a merged sensor system has to be on an equal intelligent level as the human's accessible database, i.e., the brain. To get a true symbiotic effect between an individual and an adjacent artificial sensor system, then it is of most importance to provide the artificial based information at a level corresponding to human context awareness, as described earlier in Fig. 2.4. This approach can be seen when strengthening or weakening the basic human perceptual opinion with additional information, even if we are provided with new and important data that is out of range for the human organs, i.e., access to new information, as illustrated in Fig. 5.1.

Sensor systems based on the context aware principle will make the information available and aware to the human consciousness hopefully at a time when not distracted by human reasoning. Also, the meaning of interaction gives the possibility to provide directives and additional information requested. When communicating context aware sensing activities to the related sensor system then, a skilled communicating structure is needed, advanced by training and teaching methods, Calinon (2007). Also in, Huntemann (2008), an illustrative application presents a high level of human-system interaction when sharing a wheelchair control system.

These qualities may also be used in an unconscious mode of a person when the context aware based sensor system recognises new and maybe important sensing information of importance to the situational awareness.

These upcoming situations may for example, be a warning from a complementary artificial olfaction system, i.e., an electronic nose identifies dangerous compounds in the close proximity of the human olfaction organ that it is not able to detect, e.g., carbon-monoxide (CO).

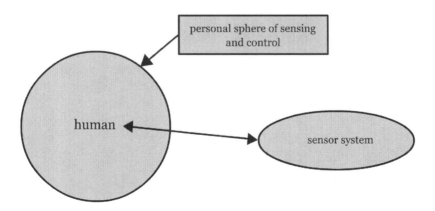

Figure 5.1. The context awareness interface.

5.1 INTRODUCTION

The traditional artificial sensor approach for the identification of qualitative parameters is illustrated in Fig. 5.2. This basic principle is that each fraction of what we consider a qualitative sensing, e.g., odour, taste or other experiences related to the human perception ability, leaves a characteristic pattern, fingerprint or response map. Based on the sensor system's operational principle, the initial process begins by establishing a measurable object. This first part in a measurement cycle will organise the sensors, i.e., refresh, clean or other measures taken to make the sensors able and ready to take the measurement. Then the measured object is attached to the sensor elements, i.e., by a chamber or finding a proper time occasion. The measurement is done at a proper time stamp to ensure that a common comprehensive picture is taken at the same moment. The signal responses collected for each sensor will establish a number of single spot measurements which convert the chemical, biological or physical reaction in the chosen environment into a joint and cooperative value of correspondent electrical parameter. Since there is an expectation, that the response from each single sensor is able to exhibit one characteristic response for only a single reaction, some sensors may naturally exhibit a response profile that is similar for several different responses, i.e., the concept of sensor selectivity. The degree of selectivity in the sensor array and the type of environment sensing responses that can be detected, are highly dependent on the choice and numbers of selected sensors in the sensor array. In some special applications, the mounting and place of each sensor can be of interest, mainly due to the application and the way of applying compounds on the array. The process, however, is dependent on the circumstances that each sensor response will be a fraction of the overall result. The result is then highly dependent on the single sensor contribution that is working in its normal condition, without exceptional variations in temperature, irregular application of compound or flow,

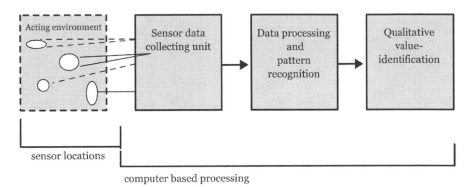

Figure 5.2. A principal quality identification system, illustrating that the sensors may be directly active in its environment... or at a remote distance and detecting different parts of the object.

e.g., in a measuring chamber. Also possible unbalanced sensor response time will affect the measurement performance.

5.1.1 Traditional Sensor Fusion Models

Traditionally, a fusion process involves a hierarchical model that considers the sensors and the sensor data as providing input values and delivering output results, for example an identification. The details of how the process between the input data and the resulting identification is formulated has in time varied among different proposed model approaches. The JDL-model, Hall (2001), is one of the most well-known models that is based on a general approach to design its methodology. The model was later improved by a fifth level user refinements, Blasch (2002). Although this model proved advantageous to find a homogenous concept within the field of sensor fusion, its shortcoming resides in the fact, that in many applications, more specificity are needed, e.g., use of extensive feedback and a dynamic knowledge base, Biel (2002). Actually specific considerations may be taken when designing a complex multi-sensor system:

The system requirement may involve parameters such as,

— accuracy achieved,
— resources needed,
— time required,
— flexibility of system, and
— cost factors.

The method or strategy involves sensor modeling and fusion methods in,

— measuring strategy,
— association of data,
— selection and extraction of features,
— validation of results, and
— of course, an extensive feedback function.

An active human related interaction can be introduced in a model and functions as an effective feedback that recognises earlier experience implemented phase of intelligence. By including the human function in an intelligent model, there is an opportunity to enable an active perception model. In an artificial sensor application, where the consideration is focussed on a symbiosis effect, the perception process may contain both a human-like passive and active process. The premises for passive systems, then the function lacks the ability to manage a mission and to act in the environment. On the other hand, by using an active system there is a functionality that is able to momentarily change the initial mission according to the occurrence in the environment, and thus redirect the sensors to areas of interest or refine the common goal, Loutfi (2005).

In Bajcsy (1985), two methods are described concerning the concept of an active perception system, the top-down and bottom-up principle. In the top-down methodology, the environment is totally unknown to the system and the system has to use accessible sensor information to perceive the surroundings and complement with existing knowledge. The system will then literally "open its eyes and report what it sees". On the other hand, in the bottom-up principle, the system will search or follow a certain given and pre-set goal in the environment.

The active perception process is mainly used for focussing the attention on specific events and field of view. In the literature, active artificial perception processes can be found, e.g., in Biel (2002a), in specific vision system, Biel (2000), odour systems Lindquist (2004), tactile systems, Robertsson (2006), and combinations thereof, Bergstrom (1998).

The other aspect of multi-sensor fusion is in its non-selective approach similar in data response and provides a co-operation of measurement in the same field of measurements, that is the sensors work co-operatively by observing the same properties of sensing interest, Lundh (2009), as illustrated in Fig. 5.3.

The proposed sensor fusion concept in Fig. 5.4 can be considered as a design structure in three dimensions, aiming to achieve different strategies in the information processing. The fusion process may of course vary between fusion applications and expected performance. The proposed organisation structure below indicates three general fusion levels in a hierarchical design. The data fusion, feature fusion and information processes can be executed at a low, medium and upper level of the hierarchical design

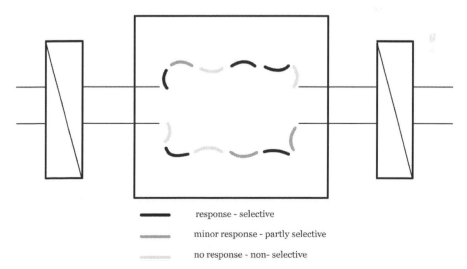

——— response - selective

——— minor response - partly selective

——— no response - non- selective

Figure 5.3. A non-selective odour identification system, illustrating that many sensors' output parameters are overlapping, exhibiting highly redundant results.

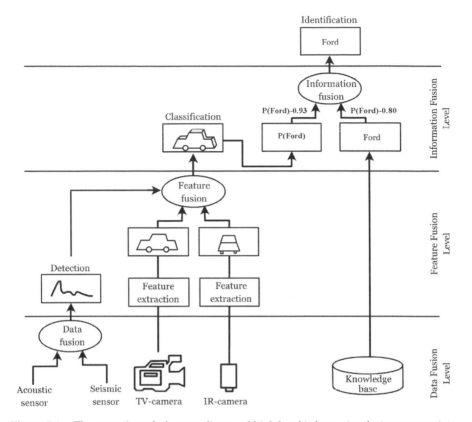

Figure 5.4. The execution of a low, medium and high level information fusion approach in a hierarchical structure.

There is typically a basic connection between the three levels. Depending on the outcome of the fusion system, after all it is of importance to meet the predefined goal, with all the means at disposal.

Outgoing from a statement, a modified definition of fusion at different accessible levels is proposed as:

"Sensor fusion/data fusion/feature fusion/information fusion is the process of combining sensors/data/features/information in such a way that the result provides more added valued information, compared to handling each source separately".

Based on the motivation, a modified hierarchical structure from Fig. 5.3 containing four different levels are illustrated below, in Fig. 5.5. The fourth structure is a redundant fusion level, where the same parameter is measured and fused with different type of sensors, e.g., all sensors are measuring a temperature parameter.

By adding an active perception process into a model, the performance will ensure the possibility to obtain an increased performance in certain applications.

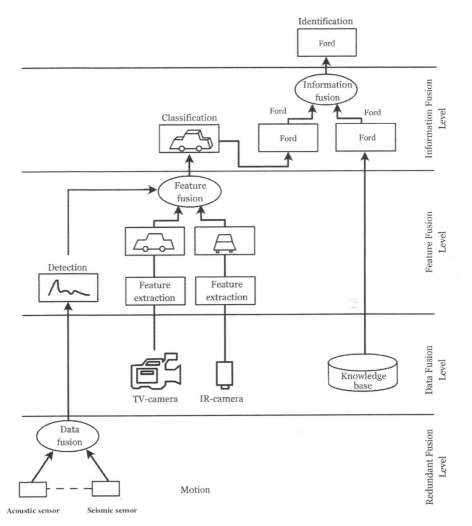

Figure 5.5. The execution of a four different fusion level information fusion approach in a hierarchical structure.

In the next section, the extension is directed towards a human related model approach involving both a passive and an active perception principle.

5.2 ARTIFICIAL PERCEPTUAL METHODOLOGY

The request for more complex and advanced sensor development has created a continuous need for better techniques in interfacing with the human ability, especially with regard to the communication of perceived information. The current limitations in human ability makes artificial sensors and computer systems that arise from the sensor interfacing inadequately and unbalanced properties may result in a conflict with natural human behavior. There are a number of aspects related to the human-based sensor approach, such as for example:

- How to make selective sensing
- How to perceive the sensor data
- How to extract important features
- How to make intelligent decisions
- How to exchange essential information

In the first aspect, the information from the environment is often imprecise and in some sense limited, often with uncertainties. Therefore, the performance of the sensor system needs to reflect a correct view of the environmental picture. A major component in designing human-based sensor systems is the process of merging sensor data into relevant information. The retrieval of the pictorial view collected in the environment has to transform into an artificial domain given that the specific qualities in demand have to be sensed and reproduced. The knowledge and earlier experience concerning the essential features of interest in an environmental picture are needed to add a flavour to the final information given and these have to be integrated in the extraction process.

Further, the question may arise on how to perform the system strategy to make relevant decisions that are based on the dynamic sensor data, followed by the process of identifying essential information. Finally, the aspect of exchange of essential information with, e.g., the human, is a crucial concern in conveying the depicted view of the environment, i.e., the human–artificial system interface. This interfacing process often controls the effectiveness of the complete human-based sensor system and when the exchange relies on human behaviour, it could bring about trust and understanding of the system's performance. When increasing the "intelligence" in a perception-based artificial system, an effective user-related interface will, in that connection, strengthen the interaction between a person (or many persons) and the system (active perception), or when considering complex environments, by the use of a high level of perceptual related interactions.

It is necessary to understand the possible features of a perceptual-based artificial system and the output information that is extracted from the measurement process. The understanding is needed to effectively make use of the information from the sensors and to contribute to a higher level of knowledge. To describe the benefits of the proposed perception model approach, a structural concept is presented below in Fig. 5.7. The perception model process illustrated below intends to demonstrate the combination effects that can be found in human perspectives of merging perceptual information and a sensor system's approach complementing the human abilities.

The methodology in artificial perceptual systems is complex, because it takes into consideration all the various aspects that are necessary in order to achieve the goal. The goal and the strategic expectations, compressed by the individual when interacting with the system, is individually-based and considered highly subjec-

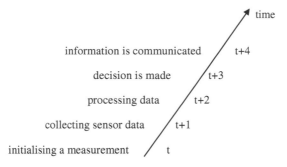

Figure 5.6. The time flow of a measurement cycle, showing the different time slots needed to secure the required procedure for each section.

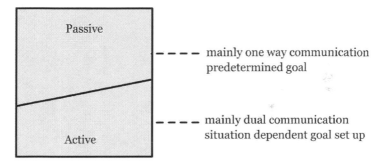

Figure 5.7. The perceptual model approach showing the active/passive approach.

tive. The outcome of the system's expectation in performance and guidelines is different depending on whether an individual needs complementing information when participating in a sports activity, than when the person requiring guidelines when crossing a dense traffic road.

The artificial perceptual process can be described as a general approach, where one or a number of sensors makes spot measurements in the environment, often at a distance away from the user of the received information. The time related collected data is often local that provides a single spot of data. Many single spot data are built together into a "global perceptual picture", that provide an information chart of the required information. An important aspect of the system performance is the communication of information to the user both from the global and local point of view concerning content and time. This aspect is of critical importance when estimating a "real time" measurement process. This is illustrated in Fig. 5.6 and indicates the significance in time relations between the individual sensor data.

When focussing on the advantages of the complete process of human supported sensor systems, then there is a need for extended efficiency in the techniques for interfacing of the different sections in a measurement cycle. The efficiency has been built upon the assumption that the best effort is made to minimise the time between the slots but without jeopardising each section's ability to fulfil its obligation, i.e., to process the activity required.

The first impression, regarding information from the environment is often of imprecise and highly limited quality. A vital aspect is therefore to obtain necessary knowledge about these often new situations in the related process. The need to know how to merge sensors together to complement each other and how to fuse the data from each sensor into relevant and requested information is obvious. The communication is a vital part of the process in order to know that the system delivers the right information at the right time as illustrated in Fig. 5.6 below. An important factor is, of course, to communicate the information in such a way that the receiver is able to assimilate the expected data in an effective format. The communication part is, as mentioned earlier, a crucial process in an artificial and perceptual sensor system which could bring an effective performance to the interaction. This process may increase the intelligence of perceptual systems by strengthening the interaction and integrating the individual involvement in perceptual processing.

5.3 A PERCEPTUAL MODEL

It is highly important to know the intended goal of the specific perceptual system, in order to create an effective task procedure. This procedure can be performed in a predetermined passive perspective or as a highly interacting phase with active capability and iterative behaviour. A procedure, where the sensor or sensors, in a system acts in a passive mode and only receives information as an observer, is considered as a passive perception system. The system then is used without active cognition, given that the overall system does not provide any feedback, i.e., mainly single communication from the environment to the system is performed. This passive type of system is also considered to be vulnerable, since any uncertainty or unexpected disturbances that occur in the environment, results in a predefined goal will most likely fail or be misapprehended. Passive sensor system applications are usually used when we want to follow a course of events in the environment, in order to establish a "state of the picture" of an environmental occurrence. The system does not act with the dynamics of the surroundings and this mode may be an effective system for highlighting and detecting changes in the environment when measuring a "picture".

In an example shown later in this chapter, a procedure where the sensor system is only sensing and communicating relevant information, is demonstrated in a passive perceptual mode, i.e., not providing any dynamic feedback to the situational context.

On the other hand, an active perceptual system is focussed on the interaction between the cognitive unit and the occurrence in the environment. Identification, re-calibration and change of focus with respect to the behavior of the system can be performed with the purpose of achieving the goals set. This principle provides a cognitive communication with the sensors, in order to achieve an optimal goal and to receive the best available information.

The citation below, given in Bajcsy (1988), depicts an illustrative description of the intention connected to active perceptual systems. However, in this citation with reference to the visual sense in general;

we do not just see, we look

makes the conclusion that a corresponding passive perception sensor system then can be illustrated as;

we do just see

In Bothe (1999), an illustrative example of the principles of an active system shows the benefits of an active perceptual system by fusing auditory and visual information. Two perceptual sensor types, a microphone set and a camera are used to visually localise and follow a sound source, carried around a room by a person. The auditory-vision sensor fusion approach illustrates the theory of merging two human related sensing abilities in a corresponding fused picture. Other examples of active perception-based models have been presented in a variety of applications, e.g., fire detection, food and water quality assessment, Biel (2002a) and Iliev (2006), with successful results.

During the last decade, a number of perception related artificial models have been proposed at various levels of fusion as described earlier in Section 5.1.1. The characteristic properties and the primary goal is clearly to provide a human in-volved solution when fusing artificial sensor information. By those means a sys-tem designer has also to consider, not only the human participation but also to measure human-like perceptual properties.

The generic perceptual model proposed below has been developed and tested in different situations. The perceptual model is split in different sub-processes into separate units, each of them corresponding to a specific task, e.g., sensing, process-ing and decision-making. The model has been successfully tested in a variety of applications, for example, in a perceptual sensing device, comprising five separate artificial sensors jointly integrated in a perceptual head device, Wide (2000).

The artificial based perceptual conception is often compared with the func-tionality of the human brain and has in this book only been intended to be used as illustrative views of related functions. The biological outstanding ability, with its highly specified performance will not be comparable to the artificial analogue sen-sor systems. Therefore some sub-processes and functions are biologically inspired, however, not restricted to these expected boundaries. Maybe the future will pro-vide us with tighter biological-artificial solutions that will complement each other in a blended integration of performance.

A comprehensive view describing the proposed generic sensor-based percep-tion model with human related abilities, as shown in Fig. 5.8, is also given in Wide (2007), concerning the background and intentions, in Loutfi (2005), the given con-cept of the design.

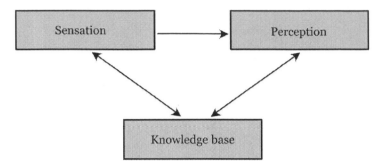

Figure 5.8. Description of the interactions in a generic sensor-based perception model with a knowledge-based unit.

The sensor based perception model in Fig. 5.8 can also be seen from the perspective of interfacing between modules or sub-processes. The advantage of a well-defined interface structure is of interest both in the sense of reusability and for defining input/output criteria of each module or sub-process. The basic design is aimed to view the human correlation and to strengthen the persuasive ideas about effective perceptual sensor systems that are able to make powerful interactions.

In the interface between the model and the environment, a main objective is to define the necessary sensors to obtain enough information from the area of interest. Secondly, a request to focus on a certain object or event in the environment will have the benefit to control and direct the sensors to an active area of events that improve the ability to reach the goal of the mission, i.e., active perception.

In the procedure between the sub-activities sensation and perception, data is arranged in a dimensional reduced organisation or in a filtered procedure and placed in a structure, i.e., a data array, in order to handle both the separated data as well as the overall time-tagged information package.

The interaction between the perception process, which is considered to be passive per se, and its active counterpart, is a double directed communication structure. The optional communication process directs the perceptual capability in handling and organising the feature selection. This is suitable for the choosing of measurement policy when deciding the structural perceptual procedure of the sensor data collection in an environmental picture. The measurement policy is considered to be of central activity in the model approach.

An important feature in the model is the knowledge-base process, which acts as an overall involvement unit in the model. The interface to the knowledge base is double directed, indicating that information gained from the process is continuously updating knowledge of the system. Knowledge gained earlier can then be obtained as additional information in future decision-making situations. This quality will of course then increase the system performance.

In the other direction, the information provided to the different sub-procedures is a powerful quality that affects the interaction between the different units in the proposed model. When operating in the run mode, the knowledge base may be directive in the following situations, as for example:

- *a priori* information, important information already known.
- run time information, dynamic changes to achieve the goal.
- restrictions or limitations, arising in the measurement process.

The proposed interface approach is considered to be straightforward with a minimum of built-in conflicts, as described above. This model approach as shown in Fig. 5.8 can be refined as a modification of the basic structure according to the previous picture, and illustrated in Fig. 5.9. The interfacing processes are defined and human relation involved in the system has been considered.

The above perceptual model approach can be seen from different behavioural perspectives and may be related to fields in biology, computer science or engineering when relating to the artificial perceptual sensor system perspective. The three following angles of approach are concerned when providing the view of a perceptual model comprised in an artificial sensor system. This aspect of the model can be illustrated in the following three perspectives:

- a human-related perspective.
- an artificial perceptual sensing systems.
- a computational perspective.

The sub-processes of the perceptual model approach can also be related to a technology point of view. The illustrative aspects are structured in general terms and described in further detail below:

- *Input interface*: a *sensational procedure* with similarities to human sensation and preprocessing activities, through a number of nuclei, of sensory information to

Sensor based perception model

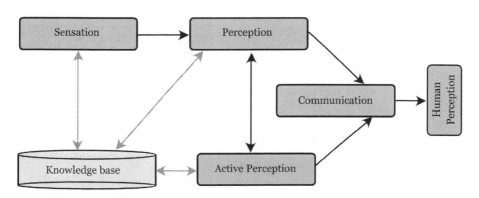

Figure 5.9. The proposed interface structure of the sensor-based perception model with human related abilities.

the brain. A number of different sensor capabilities ensure the ability to connect dynamic activities in the environment that correspond to the "attention" of the artificial system.

- *Signal analysis*: a *perceptual procedure* that organizes the received data into a "structural picture" and extracts the requested information. This process has similarities with the functionality in the thalamus and cerebral cortex.

- *Decision-making*: an *active/passive perception procedure* that handles decision-making in the system similarly to the motor cortex. In the case of artificial systems, the interpretation is performed by implementing sensing as data acquisition added with the use of pattern recognition.

- *Knowledge base*: contains all the *knowledge and earlier experience* as compressed in an artificial database unit, by similar functions to a brain process.

- *Communication*: *communicates the information* to the human user in an effective and perceptive process. Also the information may be used as decision-making by other artificial systems or stored in a knowledge base.

- *Human-computer interface*: provide the added value for the individual when *integrating the artificial information to human perception*. The interaction with an artificial system and performed by human sensing, that may be involved in the interface process. This will be discussed in more detail later in the application part.

The perceptual model approach is structured in a rational organisation built upon the human perception. However, the similarities are more on the organisational level than trying to mimic the human functions. This can be seen as advantageous when both the biological, e.g., the human and artificial, e.g., perceptual sensor systems, are intended to closely cooperate with each other in a sophisticated way.

5.3.1 Artificial Perception Model

The conceptual model concept, earlier described in Section 5.3, has been modified in Robertsson (2007), as shown in Fig. 5.10. This modified concept of a general perception model was originally adapted to the use of an electronic tongue, i.e., gustatory sensor system. In Fig. 5.10 below, the presented model is modified by including a possibility to implement the active or passive approach. Further, an added feature quality is representing the set of the environmental picture, which relates to the collected amount of measured data. The operational principle gained by using a cognition level, when focussing on *a priori* definition of the feature set, may be an attractive way in choosing the strategic approach of the method. Further it is an advantage when shifting attention between objects or events of interest in the active perception process mode. The arrows in between blocks in the figure below are not representing the specific order of information flow, but a possible interaction path. In the figure, the ability to learn from the process and explore features is represented by the direct connecting arrow (dotted arrow) and specifically

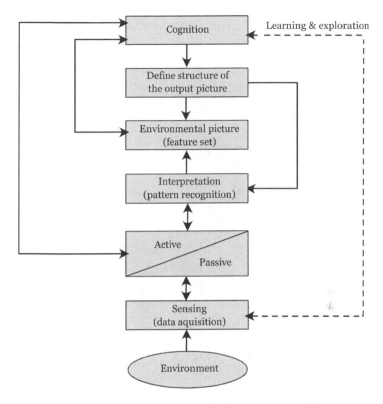

Figure 5.10. A proposed artificial perception model.

focussed on the collection of training data and evaluation of exploratory analysing methods.

The general structure is further developed to explore a passive perception model and put into a context of a gustatory sensor system, for example, an electronic tongue working as a predefined warning-based system. Outgoing from the generic artificial perception model presented above, a passive approach is proposed in the same reference.

As in all measurement systems, a useful and well-defined methodology is required. The system goal has in this approach been defined, where existing limitations and other strategic matters have been identified. In the passive mode, however, the cognitive level specifies the goal by establishing the structure of the environmental picture by defining the feature set. Depending on the specific application, the feature set can be arranged in the scale from plain raw data to an advanced analysing strategy. The sensing strategy is then planned to passively observe the environment and to measure the parameters of interest. The processed measurement is brought up to the cognition level and displayed to the user as shown in Fig. 5.11.

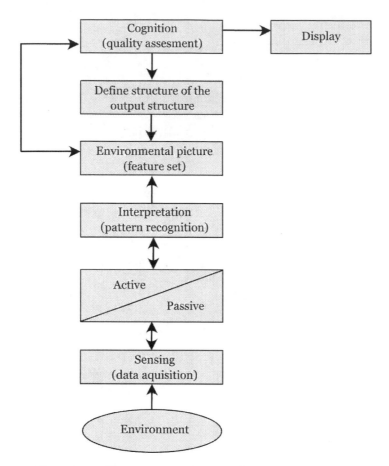

Figure 5.11. The passive perception artificial perception model.

The artificial model describing the perceptual sensor system approach aims to perform in a similar manner as its corresponding biological perception models, although not in an identical behaviour. The human-like artificial structure has its limitations and the cognition level cannot with any distinguished sense be compared to the human brain capacity. Nevertheless, similarities in features and basic qualities may form a methodology in a manner that features an useful technique in a broad sense and based upon the biological structure of perception. Further more, similarities in the concept of information flow from sensation to decision-making can also be applicable in an artificial system.

The previous model approaches have been modified and suggested to provide clarity that focus on the main stream of a dynamic information flow. This is performed with the goal to achieve a virtual, but still realistic picture, comprising the reflected and measured spot of the environment that is of interest, i.e., an

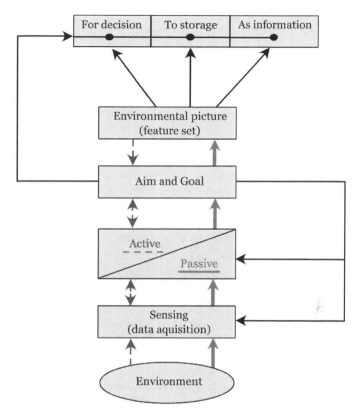

Figure 5.12. Modified artificial perception based on a goal oriented information flow model.

extensive environmental picture. The overall goal in using an artificial perception system, is a challenge that probably will bring out the need for human-like structures of the more or less rough information flow provided by the system. The resulting depicted picture describes a small replica of the real spot of interest. Figure 5.12 illustrates a modification, from the proposed artificial perception model in Fig. 5.10, where the aim of a perceptual system is further considered to be an instant picture of the spotted measuring volume in combination with the degree of measured satisfaction. The model is actually identifying the aim in information flow in the system operational principle and thus can be defined as a goal oriented model for both active (dotted) and passive (bold) interface structure, as seen in Fig. 5.12.

5.3.2 The Time Aspect

One major insight when investigating the performance of an artificial sensor system is the timing between acquired data and procedures that will provide the added value to a user, e.g., an operator. The time factor is of most importance for the interfacing procedure in order to achieve control of the essential data received and to deliver the right data on time. The timing is an essential internal function in a system that is frequently described in the manufacturing industry exhibiting similar structure of parts/data to be at the right location in an expected time slot. The concept of lean production manages the benefits of a just in time delivery concept when optimising products for each other. The production concept has similarities, which can be used, in designing complex data handling processes, for more information about lean production, see e.g., Hirano (2006). In sensor systems architectures and certainly in fusion procedures of sensor data, the need for synchronisation of time dependent information, can usually be improved. The lack of an effective management organisation of data, often varying in priority and in time, may decrease the system performance, and in some disadvantageous cases even cause a contradictory information flow.

The time aspect is one of the most important features and the synchronised acquisition of sensor data and ought to be of primary interests in most applications. This is a concern when dealing with different types of sensors which exhibit a difference in conversion time. In this case, the slowest sensor will decide the time slot for the measurement process. Many situations may be optimised by effective timing and of course the right decision of when to take the "snap shot" of the measuring spot. The cause of possible discrepancies may depend on several factors. The sensor data may be directed with a time lag, the fusing procedure is time-related or the decisions are made by different "fresh" information. Time procedures in a sensor fusion system can be identified in two basic time schedules, Biel (2000).

The time schedule, as shown in Fig. 5.13, refers to a procedure that considers a general translation of the data into a common time scale. The important feature is that the time scale can relate occurring data to each other in a predetermined schedule. The structure is organised in a direct perception procedure, that transfer the data into a common time scale where the appearing sensor data is fused to specific information. This specific perceptual sensor data flow compromised in a sequence was introduced by fusion through vision and auditory experiments, Bothe (1999).

The time schedule, as shown in Fig. 5.14, illustrates the process that sequentially affects the following data as a dependent of the received sensor data is related to the sequence of sensor data appearance. The sequential information procedure is crucial to the appearance of the following information and may exhibit an active perception procedure. In this example, illustrated in Fig. 5.14, the first fusion procedure is performed before information added from other sensor systems concludes the overall perceptual information. Thereby, the fusion process may also introduce the weighting effects between different sensor systems that may provide various priorities to the output data from the systems. In Fig. 5.14 above, we consider a sequential perception procedure that as a fist sub-process fuse vision and chewing sensing data followed by a similar activity when fusing smell and taste data. The final sub-process will merge the fused results from the two

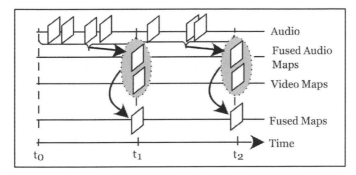

Figure 5.13. The direct perception model, where data from different sensor systems are transferred in a specific time scale.

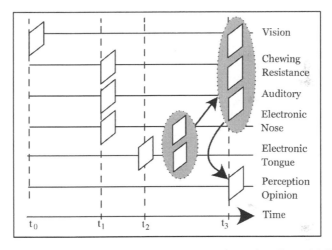

Figure 5.14. The sequential perception, where sequential time handling of different sensor systems is considered.

sub-processes into an overall information. This fusing process was introduced by Wide (1999), fusion of a complete electronic head experiment containing five basic sensing abilities, (including also an additional auditory related sensor system).

The fusing procedures above relates to a direct and sequential time representation respectively, and the models contain different structural concepts to deal with the collected data that are related to the time factor in a sensor fusion system.

An additional variant may also be included in the strategic view of a generic sensor-based perception model, presented in Fig. 5.15. This specific approach is considered when past information is revealed from storage, i.e., a memory. The procedure has then the advantage that it will consider and fuse earlier experiences from a memory in conjunction with the received sensor(s) data at specific time tags. The earlier gained information is of a historical value and is able to provide

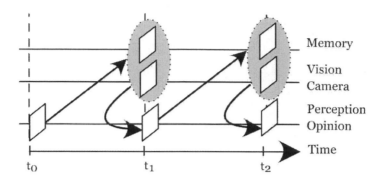

Figure 5.15. The modified direct perception process, where data from a memory capability is added to a model.

additional effect on the performed system output, when sensor data are merged with previously presented time-related opinions. In the Fig. 5.15 above, an illustration of a direct perception model with memory capabilities is described. In the model, memory-based information is merged with a vision system, i.e., a camera, and the result strengthens or weakens a momentary opinion. Sequential perception may certainly also include constantly memory capabilities, or earlier information from the memory that can be used when this knowledge is requested from the model.

The proposed time models are viewed as principal structured organisations based on a general time scaling. The main aspect is consequently to settle the synchronisation of the activities within the system. The time aspect is of an increased interest when the environment exhibits a dynamic and complex behaviour. The designing structure and the knowledge of merging the similar real-time and related data from the environment in effective short time tags, is of importance and a guiding principle in sensor fusion system design concept, Loutfi (2005).

5.4 VERIFICATION OF SYSTEM METHODOLOGY

The methodology for estimating the performance in an artificial perceptual system is of vital interest with regard to continuously achieving increased knowledge concerning the system's behaviour. As stated earlier, this type of human-related sensor systems are often related to the subjective views of the user, i.e., an interacting individual. Therefore, the systems are typically highly subjected to the dynamic complexity, mainly depending on the individual who is interacting with the system. The experiments to verify the performance of a system are then naturally dependent on the individual's reactive interactions with the information and the ability to maintain an effective decision-making capacity. A proposed verification methodology that is frequently used in the field of computer science, is noted as observation-based experiments. These experiments can perform as a joint system-operator verification, and can be a useful approach in performing tools for verifying the system's performance.

In artificial intelligence (AI), a verification methodology is proposed in Cohen (1995), where experimental results are data, often presented as quantitative representations of the system behaviour, that are further processed under consideration and tried to estimate the performance. The numerical data can then be sorted in three qualitatively related types depending on how the parameters relate to each other:

Categorial-, Ordinal- and Numerical data.

In case of artificial perceptual systems, a direction towards an adjustment is needed. Additional sets of data have to be specifically performed in proposing the following distinct types:

Categorial data: is regarded as sets of classes. For instance, the senses may record a data set as classes in vision, auditory, olfaction during a time period.

Ordinal data: can be put in order or ranked as a priority. For example, in an experiment the vision sense has provided more information than auditory and olfaction, which have a minor effect on the experimental result.

Numerical data: comprises quantitative data, for example, the numerical-values of hue, saturation or temperature of colours.

Further the three qualitative and distinct types of data classification can be extended with the following processing types:

Fusional information: is related to the fusing algorithm, that evaluate and estimate the information gained from the received data type. For example, taste lasts longer than smell in aged people.

Interface status: will be observatory experiment, indicating the impression of the added information in strengthening or weakening the individual decision-making. For example, I am able to notice that the surrounding light is very dim for this type of equipment.

Obviously, we have to take into account a traditional statistical experimental methodology for the artificial perceptual system and an observatory methodology for the human part of the overall system. The performance as shown in Fig. 5.16 is actually shown in the individual's satisfaction and the added value of the increased human capability.

In a confirmative aspect of complex experiments to verify performance of sensor systems, there is a need to observe obtained effects in the output. There are various statistical tools to calculate the requested relationships and to visualise the resulting output, as for example was mentioned in the previous chapter.

The main issue in using an effective methodology to verify the system performance is to:

— *make* related classification of the data,
— *observe* class behaviours,
— *evaluate* the analysis process, and
— *estimate* the accuracy of the presented "environmental picture".

The make, observe, evaluate and estimate objectives may in a verification phase of a system's performance be useful tasks to process. The observation experiments are necessary when a direct manipulation of data is not possible. The observation experiment by the issues: make, observe, evaluate and estimate principles, seems to answer important values in getting a credible response. However, the input situation may of course also affect the trustworthy process in validating the expected situations.

5.4.1 Human Evaluation

Human evaluation of food products contains a specific description of a regular, structural and objective procedure. The evaluation in sensory analysis is a scientific discipline that measures, analyses and interprets responses on food behaviours that are related to the human perception. This simplified definition of sensory analysis include both a quantitative as well as a qualitative approach and may also involve measurements from consumers, trained experts, and correlated artificial sensor systems, i.e., perceptual based devices. A strategic combination of these competence's create a balanced knowledge in cooperation, that in most situations will exhibit a useful and effectively composed evaluation panel to explore and verify food related products. From the human perspective, it is interesting to notice that in order to strive in further development of composing healthy and tasty products, we actually use the human sensing organs as references. In understanding the human behaviour when experiencing food, we provide human panels that use their individual sensing organs to measure properties, like appearance, smell, taste and consistency for example as illustrated in Fig. 5.16. The descriptive analysis, Gacula (1997), is a scientific area successfully establishing quality control to maintain the expected sensory quality parameters in testing of products in concordance with consumers' response. A useful reference in the used techniques in sensory evaluations can be found in Meilgaard (2007).

The specific human perception is often also used when estimating the measured properties from artificial instruments. The calibration and verification of an artificial perceptual system is complex and the response is qualitatively dependent. The experimental methodology is often compared and calibrated with human "experts" or expert panels. Expert in this sense means that the person has proven perceptual abilities and is able to put words to a sensing impression. Of course, ordinary consumers also may be involved in the panels.

Figure 5.16. Illustration of a human panel when the sensing organs are exposed in a wine testing process. Image courtesy of William Lawrence. © 2010 William Lawrence. All rights reserved.

The manually selected data obtained when testing food products may be increased by experimental design. Computer programs are able to significantly increase the possibilities of extracting useful information from the data. Initial data from the experiment design is used to test the best possible experimental plans that explain the trend of the experiments. It is, however, beyond the scope of this book to penetrate deeper into the fundamentals of experimental design. Further details on this subject can be found for example in Esbensen (2000).

5.5 THE HUMAN IN THE LOOP OF FUSION

From the previous sections, we may draw a not too controversial conclusion that in many cases the use of the human sensing organs is highly competitive in sensory analysis. The human ability, in acting as a perception unit in a complex measurement system including artificial sensor systems, is still in frequent applications convincingly competitive and in many cases an effective solution. The human participation, as a sensing unit in a measurement loop, is in many situations of importance for the overall measurement system performance, especially when humans contribute with highly complex and advanced qualitative input. In even more involving activities, human capacity is normally highly directed in the analysis and decision-making process, providing the human intelligence and sensing capabilities. This argument is often needed in complex measurements.

5.6 APPLICATIONS

Applications, where the human is interacting in the system loop can be frequently seen in daily activities. Specifically, three specific procedures can be mentioned as involved in a measurement system:

— human activities, where the individual is acting as a sensing part, in a measuring and reacting function, being an active node system in a measurement system, i.e., providing quality sensing information and acting in the environment.
— human activities, where the individual is equipped with sensor data and can provide an extended analysis and decisions, i.e., perform more adequate.
— A combination of the above functions.

These types of activities can be identified in daily activities and there is a common value to understand these premises. The performance may be improved if the awareness is considered and the human limitations are explored.

For example, the driver's involvement is quite different when driving the car in daylight contra a dark night, compensating the limited control of the situation by using all senses to percept the situation. The passenger also has a passive attitude to the driving situations. If the passenger is feeling secure, an indirect perception of the situation noticed, but if feeling insecured then this will be make the passenger more active in the driving process.

The human as a constant sensing creature is a complex system of verifying activities in confirming that the situation is under control. This specific situation makes use of every existing and performing sensing ability in the body, and may occur in daily situations when taking the elevator, preparing dinner or collecting the kids from day nursery.

It may be of interest to know our own behaviour during daily activities and to estimate the type of involved activities, according to the above list, when acting in different interactions, for example:

— when telling your friend that you have noticed that his bicycle has a flat tyre.
— when deciding to stop because a warning sign is flashing on the instrument of the car.

"If I have known, I would have done it in a different way"

is maybe an excuse, but would it really have made the person to make another decision if he had been equipped with more information?

We should experience our own behaviours during a day and try to determine each situation related to the type of involvement. The personal situation can, with advantage, use the same premises as used for the verification of artificial system performance, as stated in Section 5.4. The make, observe, evaluate and estimate objectives may also in daily situations be used as a verification phase of an individual's performance. How do we make use of data, can we make conclusions from how we observe behaviours, evaluate the perceptual process or estimate our decision strategy? These are complex questions to be answered and even more complicated to be aware of, but still the difference between human perception and complex perceptual systems are still conceptually not too different.

Chapter Six

Artificial Perceptual Sensors

6.1 INTRODUCTION

In this chapter an introduction to artificial perceptual sensors, their properties and possible applications are viewed. Also, a deeper understanding of possible structures is provided with a clear aim to generally apply artificial sensors as "extended" tools in conjunction with the human perception system. By the word "extend", the meaning is viewed as to displace the point of sensing remotely away from the human body. For example, instead of using our biological tongue to get in direct contact with food compounds, a rather attractive and convenient thought is to make use of an artificial sensor system. The system can then conveniently in an external process assess and evaluate the quality of a food, e.g., by using an electronic tongue system placed in close proximity to the body. This method surely provides a safer concept and could, at extensively used commercial versions, also provide a qualitatively more efficient evaluation by complementing the human organ. The consequence of the new gustatory and olfaction sensor device can be seen as a new generation of sensors built for an individual's remote use, that are able to provide fast and accurate human interaction including a safe approach for detection outside the body.

Humans exhibit natural properties that decreases the sensing abilities when ageing. The taste and smell functions no longer provide us with new and enriching sensations to the extent that was obvious at younger ages. The perception of touch provides us with a considerably less extensive stiff texture and exhibits only the basic impressions. Furthermore, the vision sense usually indicates that adults constantly experience a decreased performance and in combination also give rise to a constant need for additional light when reading. We wake-up one day and realise that we have to hold the newspaper at a distance of an arm's length to be able to read the text. Then when reaching the "middle age", we normally become long-sighted and need to correct our eyes with correction lenses or spectacles.

In a world where humans tend to be overloaded with momentary decisions and actions, then a need to complement human perception performance is, in many cases desirable. The ability to mimic human perceptual properties by the use of artificial sensor systems has been an area of continuous interest. Since the emergence of smaller and integrated devices in the late 20th century, that exhibited an extensive increase in performance, and used new production technologies, have apparently resulted in new applications. To date, applications that involve, for example, artificial olfaction and taste capabilities are frequently included in quality evaluation in the food industry. Also applications using vision and auditory information are now frequently used in order to complement human capabilities.

Many of these applications use, what is called electronic; nose, tongue, hearing, touch and vision sensors — that effectively work in conjunction with process operators. That is to say, sensing devices of various degrees of selectivity, along with advanced pattern recognition components trained to discriminate between both simple and complex human-based sensing is an attractive solution in qualitative measurement technology. The output pattern can be seen as an "environmental picture" describing the activity of interest. Also for individual use, there are various needs to perceive information from the proximity area outside the body, to directly transform occasions by sensing many sets of complementary devices and to merge it to meaningful information.

The information perceived from a perceptual-based sensor may vary in operational principle due to the measurement ability in order to respond on certain properties. For example, physical sensors differ in the conceptual principle from chemical sensors by the ability to the range of measuring parameters. According to Nanto (2003), approximately hundred physical parameters can be measured by physical sensors compared with a range of several orders of magnitudes larger for chemical sensors. The development of advanced chemical-based solutions will most likely emerge in the near future, since there is an urgent need to also detect and identify the huge number of existing chemicals, for example as in the transportation sector illustrated in Fig. 6. 1.

Figure 6.1. Transportation of chemical compounds is frequently performed that exihibit increased risk. This makes it more important to control each mode of transportation. Photo courtesy and copyright Peter Wide © 2010.

6.2 PERCEPTUAL-BASED SENSORS

The most frequently used sensors of today are indeed based on physical properties to convert the inquired quantity. The physical-based sensor techniques have emerged from basic demands of the industrial era since years back. Nowadays physical-based sensors are frequently used in many different types of sensor techniques that indirectly convert environment properties to corresponding electrical sensor data ready to interface with computers. For example, several thousands of sensors are strategically placed in an industrial paper production line, where they are incorporated in parallel systems to be able to confirm the normal process performance status and indicate for deviation in expected behaviour. There is certainly a clear fact that considers physical sensors to be a main base for the advanced industrial progress which had taken place during the last century. In this section we will further discuss the perceptual-based sensor systems.

6.2.1 Auditory-Based Physical Sensors

Human beings have a primarily dominant structure in the visual oriented perception. Compared with the visual world, the sensations of other senses, e.g., tactile or auditory, are significantly less developed in our lives, Norretranders (1998). Correspondingly, we may recognise concepts that are described in a visually dominant manner and descriptions that are primarily assumed from visual objects, Blauert (1996). This peculiarity may be seen in expressions where visual object related functions experienced as dominant, for example, we usually state that "the phone sounds" instead of saying "the sound phones".

The artificial sensors that complement the human perception are today a natural element in the individual striving to freedom and an including way of living. The development has facilitated the available techniques to reduce individual acuity and has been since generations using implants and correction aids as a mean to enhance our own sensing. These aids have facilitated the acuity for many people and made them experience a normal life.

The aids shown in Fig. 6.2 illustrate an example of the auditory developments of physical-based sensors that are used as complementary aid. Through the years the supportive devices must surely be the most eagerly awaited benefit for the needy individual and would indeed have provided a better live. The spatial hearing experience is nowadays an interest of intensive research, where the challenge is a system's ability to control and in some manner also optimise the attributes of an auditory sensation at a distance from the human. The spatial hearing is considering the listener at a distance and certain directions from the sound source to correspond as closely as possible to the conditions at the source. This phenomenon will often create a hearing to many people who are around due to the often miserable audio-related conditions in a social interaction, in group-mingling as well as when communicating by using mobile phones as communication between two individuals. The phase dependence of auditory signals also influence on the information

 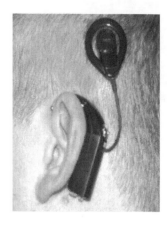

Figure 6.2. (Left) Auditory amplification as passive ear-trumpet type. Image courtesy of Alex Peck Medical Antiques © Alex Peck Medical Antiques 2010. All rights reserved. (Right) Active cochlear implant system.

received by both ears and is according to Blauert (1996), the far most important physical parameters of spatial hearing. The concept of spatial hearing embraces the relationships between locations of auditory events in conjunction with other parameters, e.g., time, space, and particularly those of specific sound events, as well as others such as those related to the physiology of the brain.

The auditory perception is an excellent example, where active technology solutions may create new possibilities for people with impaired hearing or simply have problems to participate in ordinary discussions. It also demands technology development to design modern audio directed devices that may separate ordinary speech from unwanted disturbances, i.e., surrounding noise.

6.2.2 Visual-Based Physical Sensors

The visual perception ability is, as stated earlier, nowadays an increased problem for a huge part of the population. The inability to visually sense the environment is a huge limitation that radically limits people's perception and restricts their activities. An eye correction, by external spectacles as shown in Fig. 6.3, is indeed a powerful complement that have helped many individuals for generations to adjust the defect of the visual acuity and by a moderate addition improve the result by providing a more effective perception. In the industry, automatically darkened glass mounted in welding helmets is frequently used to safely protect the operator's eyes during the welding process. Probably we will experience an emerging interest in new optical functional materials and structures for LCD-technology. Perhaps the technology will be implemented in ordinary sunglasses for elderly people, having a general problem with a strong sunlight that may automatically adjust their spectacles to the users performance when moving from indoor to outdoor in a smooth transitionadapted to the individual's preferences.

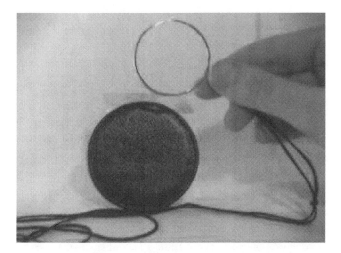

Figure 6.3. Visual acuity correction aid, glass lenses as a human complementary device. Image courtesy of www.optiker-holz.de. © 2010 www.optiker-holz.de. All rights reserved.

In contrast to the advanced sensors systems of today, compared to the simplicity of optical means to correct the eye lenses, we have nowadays gained a technology development that are able to detect long distance behaviours. Camera systems can visually be in a remote place at a distance from the body and the substantially increase the perceptual performance of the eyes. This revolutionary technique may not provide a general increased performance, but truly complements the human visual ability in its neighbourhood. In the following section, two illustrative examples that demonstrate the value of complementing the human visual sensing system, will be given. The examples are taken from a general situation where the human vision alone is not able to cope with effective perception ability.

When travelling faster than earlier generations before has adapted into its natural habitat as illustrated in Fig. 6.4, we simply cannot expect that our perception will be able to cope with various situations, e.g., when driving a car at high speeds, for example higher than 100 km/hour. The driver has to divide her driving attention between attention e.g., manoeuvring driving functions or comfort functions as the music player. Further, in such situations there is a world outside the car that the driver has to relate to and manage. The rate of environment situations change frequently and is challenging, e.g., other traffic, lights, roads and lanes. The driving situation will indeed give rise to the view of the driver's capability and how much information a human may cope with in high speeds when managing arising and changing situations. If a person has to cope with the environment but when two occational targets happen in parallell then each of them requires to be handled with a separate response. Furthermore, if the two targets in the environment are detected in a rapid succession, then the response for each, or both of them are delayed because the brain is only able to initiate one action to each of them in a

Figure 6.4. The degenerated perception ability in a traffic situation at dawn when it is raining. Photo courtesy and copyright Peter Wide © 2010.

sequence process. It has been estimated that the fastest speed to process divided attention without errors may be as low as two rapid and sequential targets per second, Glass (1986). If a person's vision system is able to update the visual scene at that speed, everyone may calculate the driver's capability between every update, i.e., the distance the car moves between two visual updates. Furthermore, if there is an aged driver, the calculated distance will most likely exhibit a larger value.

The second example illustrates the complement to the human vision by an artificial night sensitive system. Lately, night vision systems have been of interest to the car manufacturer. A night vision system built in the front of a car may indicate to the driver that the car is out of its driving path or that an obstacle in form of an animal or a human may be in the collision direction with the car. The effectiveness is as always a communication issue, indicating the need of adapting to the human user's perception ability. When driving the car in high speed, the warning system indeed need to communicate a possible danger to the driver in an effective manner that is adapted to human capability.

6.2.3 Tactile-Based Physical Sensors

Artificial tactile-based sensing development has emphasised on finding physical object shapes and contact forces. However, there are other tactile functions that are of important features, in for example an artificial hand concept. Tactile texture recognition is an essential function that refers to the qualities of the physical object surface. The knowledge about surface and surface texture, temperature and frictions are important properties for estimating an object and to recognise its physical properties. When a person moves her fingers across the surface of an object, the complex feeling apart from its shape and (a)symmetry is by tactile-based sensing recognised as the person's perceptual experience of the object's texture. The haptic perception is naturally integrated with the visual information received, Ballesteros (2005). Earlier experience about surface features is in this concept, a main subject for the learning process. The process of mimicking the human properties is a natural development aspect when modelling artificial sensor systems, Shirado (2005). Tactile sensing has gained an increased interest for example in robotic manipulation, Tegin (2005).

The properties of tactile texture functions refer to the immediate and tangible feeling of an object's surface. Precise and sensitive artificial sensor systems have to provide qualitative texture sensing values and involve according to, Mukaibo, (2005), three physical properties — roughness, softness, and friction. The three texture properties are also known to constitute texture perception of humans. The sensors are directed to measure the three specific types of information when adopting the mechanism of human texture perception. The use of multi-fingered artificial hands are especially interesting when mimicing the human hand function in a five-fingered hand, as a performance in prosthetics, or humanoid service robots. An artificial hand may also find its application in remote control and teleoperation applications. In L. Robertsson *et al.* (2007), a learning procedure as shown in Fig. 6.5 is proposed when learning the grasping of primitives by an anthropomorphic robotic hand and by using the learning-by-demonstration technique.

The goal, with an increased human value, is to provide an effective functional similarity and behaviour between artificial and real limbs that feel convenient in e.g., a hand prosthesis. The direction in research is indeed challenging when aim-

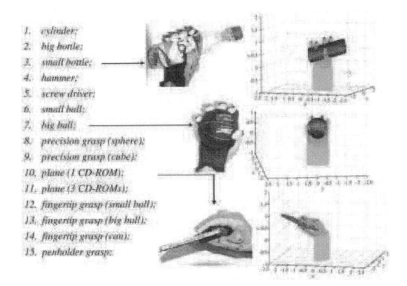

1. *cylinder;*
2. *big bottle;*
3. *small bottle;*
4. *hammer;*
5. *screw driver;*
6. *small ball;*
7. *big ball;*
8. *precision grasp (sphere);*
9. *precision grasp (cube);*
10. *plane (1 CD-ROM);*
11. *plane (3 CD-ROMs);*
12. *fingertip grasp (small ball);*
13. *fingertip grasp (big ball);*
14. *fingertip grasp (can);*
15. *penholder grasp;*

Figure 6.5. The dexterious hand used in L Robertsson *et al.* (2007). Reprinted from Int J Human-Computer Studies, Volume 65, Issue 5, L Robertsson *et al.*, Perception modeling for human-like artificial sensor systems. Copyright (2007) with permission from Elsevier.

ing to establish an improved interface between the human and artificial prosthesis, that exhibit advanced skills in brain control and fine-tuned grip functions that correspond to a normal hand function. The prosthesis development aims to make use of a complete system approach, when mimicking human behaviours in order to provide sensory and motor capability to make fine-tuned grip functions. When able to be controlled by brain thoughts then we may experience a new technology era in the field of prosthesis, Sebelius (2005).

6.3 PERCEPTUAL-BASED CHEMICAL SENSORS

A chemical sensor is a device that transforms the chemical state in the measured sensing volume in a chamber, or area of a surface, to suitable detectable data. The perceptual-based chemical sensors, have a direct additive effect on the human-based principles and are useful to complement the olfaction and gustatory sensing of humans. However, their operational principles are quite different and not quite competitive with the biological sensing abilities.

The liquid-based sensors often utilise an electrochemical principle using the chemically influenced response of an applied electrical signal. Electrochemical techniques can preferably be used to measure chemicals and some biologically related compounds that are reducible (or oxidizable). In electrochemistry, the analytical techniques are typically based on oxidation/reduction reactions, and normally called redox-reactions, Skoog (2004).

The use of electrochemical-based operation principles, Nanto (2003), can be catagorised into three techniques: sensors using conductometry, potentiometry and voltammetry principles, Wang (1994).

The measurable current arises when a potential is applied between one or several individually designed electrodes and a reference electrode. The measurement procedure is related to the surrounding liquid and the device detects the current behaviour. When applying an initial pulsed current of both positive and negative levels, the control of attracting or rejecting chemical molecules located at close proximity to the electrodes can be modelled by a chemical principle of, e.g., Heimholtz double layers model. Different model approaches are described as the variation in ion concentration with opposite charge of the electrode(s) distance to the electrode surface, Holmberg (2002).

The chemical gas sensors are based on the general principle that analyte molecules get in contact with a chemical material, which is sensitive and causes a change in the properties of the surface material. The material can typically be a Metal Oxide Sensor (MOS sensor) that consists of a coated surface of a sensitive material. The surface is doped with a small amount of catalytic metal additives, e.g., palladium or platinum. The doping of the sensor element changes its operating conditions, having an effect on the sensor selectivity to different odours and compounds, Loutfi (2006), Lundström (1981).

By modifying the restricted definition on an electronic nose from, Gardner

(1999), to also include a generally adopted statement of perceptual-based chemical sensors, we end-up with the following definition:

"a *perceptual-based chemical sensor* is a device which comprises an array of electronic chemical sensors with partial specificity and an appropriate pattern recognition system, capable of recognising simple or complex *quality measures*".

An emerging interest is shown on chemical sensors, mainly due to the increased demands on monitoring and supervision of qualitative parameters, for example, personal safety, security and food detection.

The extensive field of feasible applications using chemical sensors as qualitative indicators shows that the emerging technologies will be of interest in developing new solutions, specifically for increasing the human perceptual ability.

6.3.1 Olfaction Sensors

The extensive field of gas sensors is providing different types of operational principles in the sensing device. The sensing techniques used have certain advantages related to their specific application and the choice of operation principle may then be of importance to fit the best sensing device and provide the best possible performance.

The general technique in a chemical gas sensor is based on the principle that analyte molecules get into contact with a chemical material, which is sensitive to the molecules. The properties of the sensing material change relatively to the presence and concentration of molecules.

A traditional architecture for analysing and comparing odours is built upon the measurement principle that the molecules containing complex compounds of odours will leave a characteristic fingerprint that corresponds to the measured compounds. The operational process is often included in an airtight chamber by controlling the measurable gas, inlets and outlets. The chamber is effectively controlled by the flow of measurable gas, temperature, proper reference gas and the cleaning process required. The sensors are strategically placed in the chamber to receive a proper amount of gas flow and the signal response is performed on each sensor due to the individual selectivity to each measurable gas, as illustrated previously in Fig. 5.3. The response from each individual sensor will provide a fingerprint of the measured compound. The sensor response will further be processed to achieve a specific pattern describing the quality of the gas.

The olfaction sensor is often expressed in terms of "electronic nose" sensors. This is often misleading due to the fact that the capability and measurement principle is not comparable with the structure of a nose. The background of the term is related to its capabilities and the ambitious operational principle was once to consequently mimic the human biological perception. However, the main concept is still focused on the aim to identify chemical airborne odours and different chemical mixtures. The principle is not comparable to the advanced human capability of decomposing odours into their chemical components. In fact, the olfaction sensors,

Figure 6.6. An electronic nose system mounted on a mobile robot, comprising of two "nostrils". Photo courtesy and copyright Peter Wide © 2010.

also named electronic smell sensors, cannot be fairly compared when an artificial system containing 20 to 100 sensor elements in comparison with a human capability of estimated 10 million olfaction receptor cells. The number of receptor cells in a dog's nose may be even more, perhaps 20 times, compared to a human olfaction organ, Schiffman (1996).

In Fig. 6.6, an illustrative example, Loutfi (2004), is demonstrated. A mobile robot is controlled by the input from two nostrils as built in a compact sensor system. The benefits, as seen in the figure above, can be achieved in the phase detection of airborne molecules that can detect the flow of air-driven compounds. This type of applications may be directed towards identification and localisation of airborne compounds in order to find the source of contamination, e.g., monitoring of volatile compounds emitted from food packaging board products, Forsgren (1999). A thorough overview of olfaction-based sensors is provided in Pearce (2003). Also a more extensive overview of artificial olfaction systems can be found in, Loutfi (2002), Deisingh (2004), and Perera (2002).

6.3.2 *Gustatory Sensors*

Conceptually, the electrochemistry-based sensors are focussed on the chemical response of the applied electrical signals, Wang (1994). Liquid related applications have been inspired by sensor technologies mainly based on conductivity, Chen (2005), polymer films, Guo (2005), potentiometry, Legin (2004) or voltammetry, Winquist (1997).

The principal technique of the taste sensor presented in this book is based on voltammetry. This technique uses a potential that is applied between two or more electrodes and the resulting current magnitude is measured. An electrochemical redox reaction occurs at the contact surface of the electrodes and makes the current individually-based measurements on the respective properties of the liquid. More accurately, all molecules in the proximity of the measured liquid that are electrochemically active below the applied voltage will contribute to the redox response, Lindquist (2007).

The drinking water sensor system in the figure above is in a natural commercialisation phase of the developed technique. The system will communicate a light indication depending on the tested water sample, indicating whether it is safe to drink the water and before an intake of the liquid into the mouth. The application is demonstrated in a prototype, as shown in Fig. 6.7. The system approach exhibits an external test, while the person is filling the glass and in the meantime evaluates the liquid at a proximity outside the body. Various tests have been performed, with a focus towards the quality assessments of drinking water and applications from the source to a person's intake of the liquid, Scozzari (2008).

The operational strategy of an electronic tongue approach can be viewed in the following structure when building a base concept. In the concept below, the technique of an electronic tongue uses voltammetry and the adaptation is focussed on applications of sensing water quality, e.g., in measuring wastewater in a washing machine, Olsson (2008). Of course the fine-tuning of the type, range and selec-

Figure 6.7. An example of a conceivable design of an water test. A green light is shown on the joint of the testing unit, indicating no changes in the water quality. Photo courtesy and copyright Peter Wide © 2010.

tivity of sensors increases the capability to fit specific application requirements.

— Base concept $\longrightarrow$ electronic tongue,
— Operational principle $\longrightarrow$ voltammetry,
— Application $\longrightarrow$ drinking water,
— Sensing capability $\longrightarrow$ bottled water, tap water, wells, reservoirs.

6.3.3 A Complex Operational Integration

The measuring operational principles in artificial sensing systems are complex because they take into consideration all the various knowledge of sensing the fuzzy aspects that are not very often well-defined measuring parameters, e.g., water- and air-quality comfort. Typically, one single sensing unit cannot in fact perceive all the aspects of requested specific information, which involves a qualitative measurement. But a multi-dimensional approach of the operational principles involved is necessary in order to achieve the level of requested quality parameters of interest. Typically, input values to an information providing sensor system is in a complementary form, such as, it overcomes the non-specificities on behalf of a single sensor solutions by the use of a multi-dimensional approach. By assembling arrays of sensors in a certain strategic manner, it will motivate the common approach of using the benefits of dimensionality. Sensor arrays assemble possibilities of various kinds to increase the motivation to measure and identify complex qualitative parameters. Increased dimensionality can by advantage be used, and will be further effective if also complementing sensors of other operational principles are added to the sensor array, Sundic (2000).

A large interest today is to apply chemical sensors, not least because of increased demands on environmental monitoring, food quality and safety issues. These, and similar application solutions that require small and cost effective devices capable of sensing gases and toxins in our close proximity. Therefore, it is of high interest to be able to evaluate that part of the performance, that is focussed and mainly related to repeatability and sensitivity. The continuity in system performance over time is also a question to be solved in the maintenance schedule, when changing sensing elements or units after a predetermined time period of sensor change to avoid general sensor deterioration.

The aim of designing a specific artificial sensor system and especially complex qualitative measurement is of course to provide a convincing operational methodology that exhibits a coherent understanding. Here, substantial values lie in the measurement that continuous measurements provide over time. The main question here is provided in the validation of existing and operational sensor systems. The validation of complex sensor systems with a substantial part of artificial intelligence involved will naturally conduct test experiments to verify their performance.

The verification phase is important to establish a "belief" in the system, actually that it behaves as expected. To verify the properties of a perceptual situation, can with advantage, be measured and verified by another perceptual system. An interesting emerging technology has been demonstrated by the advantage of combining two (or more) sensing techniques measuring a combined behaviour from one property into the detection and identification of the other sensing system. The "system measuring system" concept is often a way of visually monitoring the result from other perceptual systems. This methodology may be a useful approach, also to validate human-based systems and conduct performance experiments. The optical gas imaging technique gives an illustrative concept. This perspective describes an excellent mean to visualise the resulting measurement values from an olfaction system by an output imaging "picture". This approach indeed provides a more human friendly and communicative process to a person and an understanding of the result. An artificial 'olfactory' image from a chemical sensor is presented (vision-measuring-olfaction) in Lundstrom (1991).

Another variant of the combined measuring system concept can be an efficient and forward moving principle, without measuring with a gas system that may indicate a leakage as illustrated in Fig. 6.8. Instead, the leakage substances can directly be identified by optical imaging the leakage. Gas leaking from a production plant or in transportation is a complex and undesirable situation that may end-up in a catastrophe for both people and material. The technique to measure possible gas leakage by an optical gas imaging device, has been proposed by Sandsten (2004). The gas leakage can be detected and identified by an optical camera system and exhibit the imaging by gas measurement principle. A leakage at an industrial

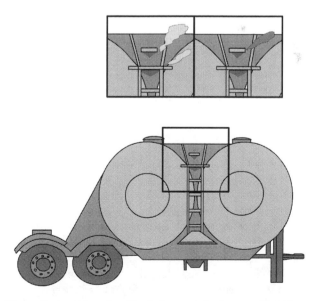

Figure 6.8. The optical gas imaging illustrates a gas leakage accident from a tank truck.

plant containing chemical compounds could be detected and visualized in colour by scanning the camera system over the industrial site. The colours are connected to the type of exposed gas.

Figure 6.8 demonstrates an illustration of the optical camera system that detects a gas leak of ethylene and ammonia as shown in the left inset. Due to the fact that the spectra overlapped, the system used a gas-correlation cell to separate out the ammonia from the ethylene with a specific gas-correlation imaging technique, as shown in the right inset, Sandsten (2004a).

6.4 APPLICATIONS

In the following application part, we will introduce several illustrative artificial sensor systems that will exemplify the influence of involvement containing an individually developed structure, which intend to increase the specificity and overall system performance. This effect will undoubtedly influence the interaction with an individual user, as well as his/her actions in the system approach when communicating with measured occurrences in a complex environment. Then, the outline of the basic concept and the modelling of an artificial sensor system are initiated, that will complement and act as a base for a creative function as additional information in a structure that is based on a decision-making foundation.

The following section will demonstrate illustrative applications that have originated from a research exploratory and innovative phase where some prototypes may still not be ready for the extensive commercialisation phase. However, these ideas have been, in controlled tests, examined through temporary implementation in industrial products or put in conjunction with individual users in solving an important measurement problem. Therefore, the time of commercialization is a bottleneck that has to be proven to present a quality and performance-based product that needs an extensive test plan and thorough examination. Hopefully, there will be several market introductions in the future that will solve expected human-based perceptual problems and add complementary information to a person's sphere of natural perception.

6.4.1 Perceptual and Complementary Systems in Industry

Sound is an excellent source of indicator that is, and has earlier, frequently been used to detect, identify and diagnosing behaviours and quality changes in industrial applications. The capacity of the human hearing as a skilled sensing unit is that it responds to various frequency shifts or added noise. It has for a long time been understood that mechanical failures, both considering initial as well as existing ongoing sound, often can be detected by the auditory and skin perception of an operator, Gescheider (1970). Illustrative examples are given that diagnose faults in rotating machines, Zio (2007), the sound shift in machine maintenance, that can be detected by both the operator and corresponding instruments.

As automation gets more and more advanced, the tendency is to minimise the human acting as a sensor. The frequency and vibration instruments are superior in performance and may provide much better result in industrial measurement applications. This aspect will of course affect the human participation in measurement situations in industrial plants, at least in countries where the labour cost is an important part of the production costs. Striving towards fully automated production plants has been on the focus for some years now, and we can already observe a state of semi-labour-automation situation. For example, in a general car manufacturing plant for mounting cars there was approximately half of labour compared with working robots, indicating also that the main task for the staff was to keep the robots working properly. The general trend of the high-tech industry has focussed more on the maintenance and control of the automated work than direct human involvement in the production. We may expect that the human still involve to confidently use her ability (in some cases still exhibiting outstanding abilities) to act and interact with the production. For some years now, the trend has been to actively search for human absence environment and to find corresponding automated or autonomous solutions. For example, in the rock drilling industry in extracting gas leakage sound from a noisy environment, Kotani (2001), and by developing a matching pursuit approach to predict small drill bit breakage, Fu (1999). Another illustrative example will demonstrate that the human skill of operator perception abilities may change to automated, and technical cost-effective mining production solution.

In the segment of rock drilling vehicles as shown in Fig. 6.9, the experience shows a trend towards unmanned autonomous equipment. Similar trends can be seen in the advanced technology production industry, where the strain of the operator in a hard environment may be stressful. A main issue is however that the drilling process still exhibits manual operations in the process of drilling in rocks. For example, the retraction of the drilling steel has after the automatically performed drilling process, to be manually handled to get it safely out of the rock. This phase is still often maintained by the operator auditory perceptual system, when hearing to the characteristic sound phenomenon appearing, indicating when the splices between the different drill steel parts are opened-up enough to allow retraction.

The sound detection ability by the human ear in noisy environments is strained and will in the long term have a destructive effect on an operator's ability. Therefore, artificial systems that substitute the human sensing factor in a production site, especially when the environment is not healthy, will always be welcomed. The paper by Bergstrand (2004) introduces a Fast Fourier Transform (FFT) solution that analyses the detected audio information to find specific behaviours. The analysing method makes it feasible to locate a specific sound related to the opening of the drill steel, following the retraction.

By artificially recording and analysing sound, there has been comparison between an implemented "electronic ear" and an experienced operator's auditory system, i.e., human ears, with sufficient results, and a commercialised system is now available. The following standard process is used in a continuous loop of

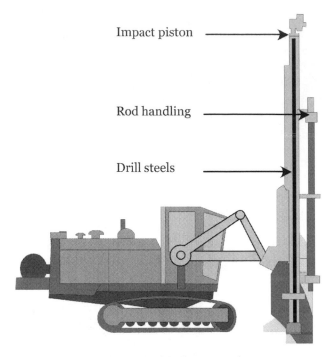

Impact piston

Rod handling

Drill steels

Figure 6.9. The vehicle used when drilling in rocks, including the drilling steel.

drilling in rocks:

- choose location,
- predetermine depth,
- stop percussion,
- and retract the drill steels.

The splices between the drill steels parts are drawn together during the drilling process and a sound detection system by human or artificial sensing identifies when it is retracted. This example shows that the human perception in a production process indeed can be executed by artificially complemented systems. Sensor systems will most likely make the process safer and time effective.

Another illustrative and technically sophisticated human-based method for identifying properties is when a loaf of bread has reached the desired temperature, in this case 96°C, and the crumb has developed a system of air bubbles, i.e., gas cells. This technology for industrial production of bread has extensive similarities with the operator's methodology for estimating the same properties. The methodology of a "finger-tip" sensing as seen in Fig. 6.10(b) refers to human-based acting and sensing, mimics a traditional method to act when the operator examines the development of the crumb simply to make the perceptual process similar to investigating the degree of ripeness for watermelons or similar kinds of fruits.

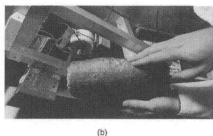

(a)　　　　　　　　　　　　　　　　(b)

Figure 6.10. The automated crumb testing system (a), compared to the human perceptual method (b). Figure (b) shows a person knocking on the bread with finger-tips. Photo courtesy and copyright Peter Wide © 2010.

The human simply puts the fruit in a close proximity to the ear and when knocking with the finger-tips at the rind, experiences the characteristic response of the sound. This frequency related human-based sensing method identifies a sound spectra that is corresponding to ripeness of the fruit. In both cases, the person is able to make a measurement based on the individual's own perception abilities. However, the "finger-tip" test is to verify that the baked bread is properly baked or the ripeness of fruits.

A corresponding artificial sensor technique has been demonstrated in Fig. 6.10 to indicate the possibility of making a similar performance to the operator's perceptual ability to recognise the status of proper crumb in the commercially baked bread process. A unit pulse is automatically introduced on the base surface of the baked bread when entering the cooling section after the oven unit. The response sound when is artificially detected and further analysed.

The artificial sensor system presented in Fig. 6.10 uses the same principles when applying a knocking on the bread surface. In Fig. 6.10(a), a detector principle using a laser distance sensor and in Fig. 6.10(b), the human ear is used to detect the response sound.

A very short duration impulse is applied to the base surface of the newly baked bread. The applied pulse will simulate a theoretical unit pulse and the response of the system is formed by the inside of the loaf bread, i.e., the crumb bounded by the volume of the crusted surface. Two alternative measurement methods have been tested to evaluate the response of an oscillating transient response function, a microphone and a laser distance method. The procedure is similar to that of an experienced operator's measurement by the sensing organs by applying a pulse and detecting the following response sound.

The industrial food production process in general, and especially in the production of making loaf breads, assume that the operator's skill is essential to achieve a satisfactory result. The industrial production exhibits complex and highly operator-dependent decisions, which are learned through experience and knowledge. When following the process of a fermented bread baking procedure, three main sub-processes are accomplished to get perfect baked bread.

- The first stage in bread making is mixing of the ingredients that are mixed together under quite ruff conditions, where both shearing and extension are involved, Hoseney (1994).
- During the next step, fermentation of the yeast produces carbon dioxide and ethanol by the fermentation of sugars, Bloksma (1988).
- The third stage is the baking in oven, in which the final expansion takes place before the dough structure changes from foam to a sponge, Hoseney (1994).

An example that is showing an artificial sensor system that evaluate the result from the baking stage and in correspondence with to the human skills, are shown in Fig. 6.10 below. The artificial system have the ability to develop effective and operator complementary solutions but also to make use of the operator skills in specific situations of the baking process.

Also, a similar human related application by physically measuring the crisp bread quality is demonstrated in Fig. 6.11, Winquist (2000). The crushing of crisp bread in a chamber is evaluated by a detected sound spectrum. Additional artificial performance is achieved by fusing the auditory sensor data with tactile and olfaction data.

Finally, in the stage describing the fermentation process, a sensor has been

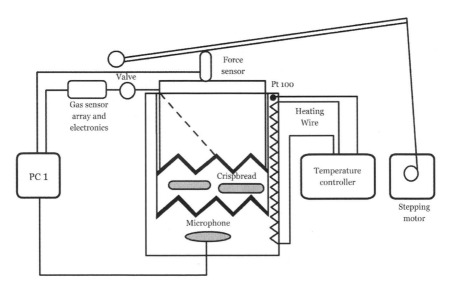

Figure 6.11. A simple method of finding the qualitative properties of crispbread by a sensor equipped chamber is illustrated.

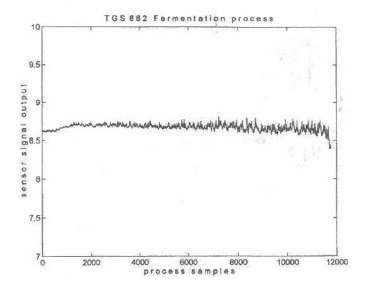

Figure 6.12. Exemplification of the response from a chemical gas sensor during a fermentation process.

implemented as a fermentation system to measure qualitative values of a known complex chemical and biological process, i.e., the bread fermentation process. Wide, (1996), shows an interesting experiment, namely that a commercial gas sensor as shown in Fig. 6.12 together with analysis of the sensor data is able to support an operator to make the optimal decision about the dynamics of a fermentation process. The signal processing is performed by time series analysis using variance, kurtosis and shewness, i.e., quantities frequently used in the field of mechanical analysing applications, deduced from the sensor data as can further be read in the paper, Wide (1996).

This technique can be compared with the traditional operator used method by gently pushing a finger on the dough surface and studying the dynamical response behaviour, i.e., the surface is with some time dependence returning to it's initial position. A corresponding method that mimics the human sensing procedure has been proposed in Wide (1995), illustrating that the same strategy is used in a sensor system, when applying compressed air pulses on the surface and detecting the reaction, e.g., with an optical distance measuring device to establish the surface time of recovery.

The mixing process is highly related to rheological aspects that change the dough properties continuously. In Wide (1999a), an illustrative experiment is presented that involves a technical-based methodology in correlation with the use of human experience by learning a system behaviour and when the specific mixing process is considered optimal.

6.4.2 Human Aid Applications

Visually impaired people have for their mobility, as tradition, been using walking sticks to detect obstacles. Despite more or less highly and complex technology solutions that are often available, still the currently most common method of range detection used by the blind is the traditional walking stick, that exhibits an effective interaction with the individual. The limitation of the walking stick is that a blind person must come into close proximity with its surroundings to determine the location of an obstacle. The development of technology aid is, despite their limitations, available to assist the blind person in moving and travelling situations, i.e., an electronic travel aid. The techniques, as described earlier also use electronic sticks that are normally ultrasound, radio frequency signals or laser to detect obstacles at a generally longer distance than the traditional stick. However, the effectiveness of the interface between a user and the system may raise some fundamental questions concerning the perceptual limitation.

As noticed, the interface between the sensor system and the individual needs to be of high importance. The interfacing device also needs to be fast and should exhibit a natural additional moment to the human perception. The intelligence of the systems available on the market is, however, moderately limited in terms of interacting with a user. For example, when producing a varying frequency of chirps (sound), vibration or pressure that is inversely proportional to the distance measured. The user often lacks additional information, i.e., if a moving object is interfering with an unexpected behaviour that may affect the users information path about the object, for example, a crossing dog. In the literature also, sensors mounted on a lightweight helmet have been found, which allow the user to obtain a reading in whichever direction the head points. This solution may not be best suited for the user, due to its inconvenient cap that most likely affects the flexibility.

The portable power source and corresponding electronic circuit in using electronic aid are often encased in a box and attached to the user's belt or fixed to the stick. The obstacle avoidance warnings are focussed on single objects and may be indicatory on distance of up to 1.5 m (12 feet), which also is a challenge for the user to master in a complex dynamic environment, Lopes (2001).

6.4.3 Human Substituting Perceptual and Interaction System

The beneficial use of artificial sensor capability is generally considered to be its ability to perform in a superior manner, when the desire in certain demanding situations may be to measure human-related properties in complicated or hazardous situations. The typical use of these sensors are in tasks where the human senses may not perform the very best. For example, a quality measuring operator who is frequently smelling and tasting jam in a production unit may exhibit a passive perception attitude and experience a mood of blasé which may result in a not optimal process of sensing.

The canary in a cage placed in a mine is perhaps the most illustrative example of biological supportive sensor systems that since years ago have protected the workers against poisonous gases, e.g., carbon monoxide. The heroic achievement is nowadays substituted by an electronic device that may provide warning in a more effective manner. The electronic device may give a warning signal before the poisonous gases reach an unhealthy level, that is at levels where the canary starts to feel uncomfortable and dizzy.

The process of developing new and uncontroversial methodologies for exploring competitive artificial sensor systems, that are able to measure and identify complex qualitative parameters, with the aim to provide an additive consensus with the corresponding human senses, may in no means be ambitious but is still a challenge. Research has, for the last two decades, been exploring an intensive perceptual-related development phase, presenting a technology "evolution" in time, outgoing from a single gas (olfaction) sensor, e.g., commercially available S_nO_2 sensors provided by the Figaro Engineering Company as seen in Fig. 6.22. This sensor and other types of sensors have made it possible to design smell sensor arrays with a selectivity range that has made it possible to detect more complex air bounded compounds. The development was perceptually followed by a tongue (gustatory) sensor system, Winquist (1997), and further expanded with an additional tactile sensor system by chewing detection, Winquist (2000). The integration benefits experience an artificial sensing ability, that have been shown in Sundic (2000) and Wide (1998). These papers investigate the fusion methods in the integration process between an artificial taste and smell system.

The concept was further pushed to make an integration of the five sensors: olfaction, gustatory, vision, auditory and kinetic parameters. In Wide (1999), an electronic head comprising the five sensing abilities was presented as shown in Fig. 6.13. The device is described as a virtual instrumentation, Wide (2001), in providing qualitative estimation and decision-making of a dynamically changing environment by combining data from different artificial sensor systems into a single set of meaningful features. The single sensor information provided is of less benefit than the aggregate of its contributing sensors.

Furthermore, the virtual feature estimation has been communicated to the operator via a computer based face — awatar, able to express the overall impression of the tested object, Loutfi (2003), as can be seen in Fig. 6.14.

The artificial head concept has been used as a demonstrating platform by using the various and optional number of perceptual sensor systems in specific experiments. Tests have been performed in a number of preferably food-related application with satisfactory results. However, after the methodology has been demonstrated in different applications, there seems to be a huge and specific task to verify the system performance and a general reliability to apply the technology in industrial applications is needed. After all, an operator seems to be more trustful in making complex and perceptual measurement tasks, even if the human aspect of verification in advanced and non-linear application seldom is elucidated.

An interesting concept is considered in the development of tactile sensing in an artificial hand, performing different gripping patterns. The industrial user and research often state that the best robot hand for picking, holding and placing

Figure 6.13. An experimental platform showing the electronic head device, where the vision cameras and mouth chamber can be seen. Reprinted with permission from Magnus Westerborn.

applications and other advanced operations can be solved with an artificial hand comprising only three fingers, e.g. Boivin (2005). The five finger hand however, has a strong impact on the prosthesis aspect, where the mimicing behaviour of a human hand is the main concern, e.g. Robertsson (2007). An illustration of a physical hand as a test platform is shown below in Fig. 6.15. Also in Fig. 6.16, the implementation of modelling the hand is viewed. Mimicking the human capability by the hand motion patterns of different grasps with effective finger movements simulated action is evaluated, in order to make a complete hand action.

The human prosthesis of a mechanical hand still has some important improvements to make. Up to date, a mechanical five finger hand with human related behaviours challenges the problem of external forces. When a single finger gets in contact with an unknown object, e.g., a cup of coffee, the contact forces have to be controlled to achieve such a balance that neither the object nor the finger is damaged. Another issue of importance is the interface between such an artificial five finger hand and the neural pathways of an user. The interface between neural signal pathways in the form of chemical impulses and the electrical signals controlling the artificial hand performance is available as a prototype electronic chip, Sibelius (2005). The promising mechanical hand concept in connection with an interface

Happiness

Left_Zygomatic_Major	1.50		Right_Zygomatic_Major	1.40
Left_Angular_Depressor	0.00		Right_Angular_Depressor	0.00
Left_ Frontalis_Inner	0.80		Right_ Frontalis_Inner	0.90
Left_Frontalis_Major	0.20		Right_Frontalis_Major	0.20
Left_Frontalis_Outer	0.10		Right_Frontalis_Outer	0.30
Left_Labi_Nasi	0.00		Right_Labi_Nasi	0.00
Left_Inner_Labi_Nasi	0.00		Right_Inner_Labi_Nasi	0.00
Left_Lateral_Corigator	0.00		Right_Lateral_Corigator	0.00
Left_Secondary_Frontals	0.00		Right_Secondary_Frontals	0.00

Figure 6.14. A model of the interfacing avatar, making the facial expressions of the tested objects, placed in the mouth. Reprinted with permission from Loutfi (2003).

chip is today built up as separate parts, but maybe, within a couple of years, make an improved system integration comprising an attractive and useful hand prosthesis. Further, the research on artificial biologically raised skin in the laboratory,

Figure 6.15. A dexterous artificial hand. Photo courtesy and copyright Peter Wide © 2010.

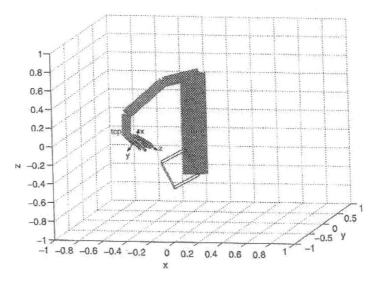

Figure 6.16. Simulation of an unknown grasping of a familiar object in a known location is demonstrated. Reprinted from Int J Human-Computer Studies, Volume 65, Issue 5, L Robertsson *et al.*, Perception modeling for human-like artificial sensor systems. Copyright (2007) with permission from Elsevier.

e.g. Kalyanaraman (2008), can be an interesting aspect of prosthesis with extended capabilities. The artificial hand will probably increase the mimicking ability and involve the skin's performance to sense the texture, temperature, etc. A major concern is also if a person using a prosthesis is able to perform better and due to more advanced technology, will be superior to the ordinary human hand? This will certainly provide other perspectives on the future abilities. Anyway, prosthesis with advanced performance will consequently, contribute to improved movements. An active and useful prosthesis in the future will most likely be able to improve the life of many disabled individuals.

6.4.4 Perspectives on Perceptual Sensing Systems

In many human-related application concepts using sophisticated measurement technology, the methods that require a strategy for collecting information about an activity is an extensive and demanding task. The kind of information required in a measurement process may rapidly change and requires a dynamic operational principle. In fact, when there is a need to get a complex situation analysis instead of the normal and often simplified measuring process of single parameters' responses with a selective-analytical approach, we may find it useful to measure general and amalgamated attributes, for example:

— quality,
— availability,

— efficiency, and
— performance.

The concepts of artificial perceptual devices have been developed and have spawned a number of challenging applications in recent years, such as, for example:

— environmental monitoring,
— public health,
— food production and
— individual safety.

Concern could be given towards the individual safety aspects, which can be implemented in all the above applications. In the following parts, we intend to go further into details about their attributes and system requirements, and introduce the concept of an artificial human sensor idea.

In this specific type of complex perceptual procedures, an overall qualitative parameter, i.e., a specific attribute, is introduced as a result of a process under monitoring, that will represent an attractive and cost effective approach for active perceptual sensing and action.

The concept of artificial and perceptual safety devices are particularly attractive since the results provide a more general qualitative information about the inquired parameters and the operational principle may not necessarily be connected to a human-related perception. In the following example, an introduction to the concept behind a gustatory device will be described in more detail. However, the general concept exhibits an organisational structure that needs to be respected and in some sense related to. The sensitivity obtained in the example below, exhibit a wide detection range that in some cases goes beyond the possibilities of the human sense of taste. In other words, this approach of artificial device may be an excellent device for complementing the performance of human perception also in applications outside the range of safety.

The design of a simple and illustrative perceptual device is demonstrated and an illustrative prototype may be easily built. However, the specific knowledge of what the system measures and how to evaluate the result, from the data point of view, is generally the advanced part of the measurement technique. In the following, the guiding principles for designing a gustatory device — an electronic tongue and an olfaction device — an electronic nose, will be specified and advised. The devices aim to provide basic understanding of the artificial human-based technology and will provide a more thorough understanding of the principles of a sensing system that easily can be extended to more advanced systems.

The proposed prototype of an electronic tongue can easily be demonstrated and built mainly by the use of ordinary home accessories. An application will be demonstrated towards the drinking water assessment and, as an example, illustrate the fundamental working principle that is oriented towards the development of a tap-water sensor for home-based applications. On the way towards an

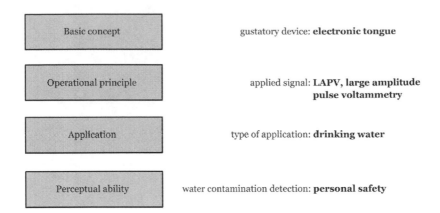

Figure 6.17. The strategy of sensor design is exemplified by the concept of an electronic tongue.

organisation path of the design there are some decisions that have to be considered. As an example, the strategy in a sensor design has to be defined and fully understood to achieve proper knowledge about the operational principle used. The chosen strategy for an electronic tongue device may be as shown in Fig. 6.17.

An illustration of the sensor design of concern is demonstrated, and the available options are hopefully known. The chosen strategy is built upon a basic electronic tongue concept applying the technique of voltammetry. Further, the sensor responses are adapted toward the application of sensing water contamination that, in this example, is illustrated by eventually identifying contaminated compounds added in the water. The fine-tuning of the chosen types of sensors is most likely not optimised and further assumption about the system performance is capable to substantially increase the capability to fit specific sensing applications is left to the reader to evaluate. The conceptual strategy is shown in Fig. 6.17.

Voltammetry is a technique of electroanalytical principles in which information about the analyte is derived from the measurement of an applied current, Christian (2004). A potential is applied at an electrode and the current flowing through the electrodes, typically working-, auxiliary- and reference-electrodes (Fig. 6.18). An electrochemical redox reaction occurs at the electrodes contact surface with the liquid of interest, which give rise to a measurable "fingerprint", that normally can be detected by identifying the specific concentration of a certain property. The sensor output is typically measured as a function of the applied potential, or simply the time. As the electrochemical method of potentiometry is also a common method used to design electronic tongue systems it ought to be mentioned. The contrast to voltammetry is that potentiometry uses a static technique, which provide the sensor system output with relative potentials that are measured. Details of different chemical methodologies can be found in Skoog (2004).

A basic component included in a general electrochemical measurement system is the electricity generating device, i.e., potentiostat, which in voltammetry injects current into an auxiliary electrode (AUX). This means that the measurement closes the current loop via a working electrode (WE) in order to impose a known difference of potential between the working and reference electrodes (REF), as shown in Fig. 6.18.

Further, in voltammetry an applied simple or advanced waveshaped voltage, v(t), is imposed across the working electrode and the reference electrode, while measuring the resulting response. The applied voltage will close the measuring loop and give rise to the loop current, i(t), which is measured.

In Fig. 6.18, a simplified schematic diagram of the experimental design is proposed. The simplified but illustrative design is normally in the extensive experiments made by a measurement cell comprising one working electrode, made of a metal that in normal devices are Au, Re, Pt, Pd. A corresponding experimental sensor is illustrated in Fig. 6.19.

However, in a simplified design for exploring the principal operation, the use of an electrode made of copper will satisfy the fundamental needs to experimentally demonstrate the function of a water safety system. As the reference electrode, or in this case, a spoon of stainless steel (e.g., 18–12) is sufficient to experience the system performance.

A simple experiment that can be easily performed is shown in Fig. 6.20. The electrodes are placed in a plastic-beaker of a tea cup size. The electrodes, in this experiment consist of a spoon and a fixed piece of copper wire. Due to chemical properties which will affect the measurement, we aim to let the same area of the electrode during the entire experiment be in contact with the liquid. Also to achieve the best possible response there may be a need to adjust the electrodes mutual distance to each other. By applying a voltage pulse sequence, for exam-

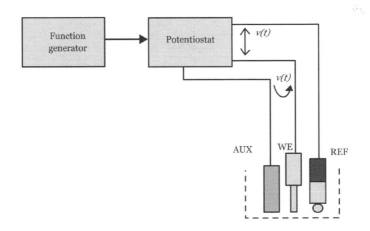

Figure 6.18. The electrochemical measurement principle shown in a simple experimental system.

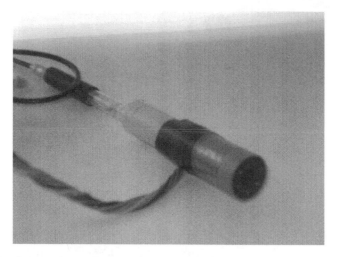

Figure 6.19. An experimental sensor containing six electrodes built-in to achieve a specific and equal area of contact to the measured liquid. The specific areas from different material can be seen at the end surface of the sensor unit. Photo courtesy and copyright Peter Wide © 2010.

ple in the range between −5 V and +5 volt between the electrodes, depending on distance between electrodes, amount of liquid, type of contamination etc. and measuring the response characteristics of the current in the loop between the same electrodes, the response will describe a fingerprint related to the specific and tested liquid. The response from the experiment can be viewed by an oscilloscope or similar monitoring device, e.g., a computer with suitable measurement programs. In case of a small contamination is added to the tested object, i.e., in this experiment on drinking water, then the expected change in the response curve will slightly change. This effect may be better illustrated if a constant measurement is performed and by focussing on monitoring the dynamical output properties. Hopefully, now there is a change in the response signal that will be monitored, and strengthened if some more or other contamination is added to the water. For example, if a small amount of contaminant from an external river or lake water is added in the experiment, that intended to simulate the contribution of dirty water, is mixed with the drinking water, then there is expected to be a change in the liquid composition that will further deflect the response signal as seen in Figs. 6.20(b) and (c).

This simple exercise demonstrates the possibility to discover contamination in the water, e.g., drinking water from the tap. This operational principle is built upon the assumption that the normal quality of the water will exhibit a stable response signal from the measuring device. Then, if a change in the response signal occurs, we may suspect that some additional compounds are identified, that will affect the response signal in a specific and predetermined manner.

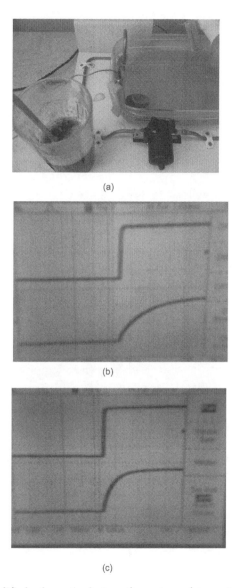

(a)

(b)

(c)

Figure 6.20. A simplified schematic design of a water safety system in (a) the movable electrodes spoon and wire can be seen mounted on the wall of the plastic beaker. Photo courtesy and copyright Peter Wide © 2010.

The proposed experiment platform, that has been discussed, may be seen as an illustrative concept to explore further possibilities, that contain functions of more advanced sensor performance. In the following part, an almost commercial device is presented for use as a tap water sensor. The system has a safety concept and is expected to be a natural part in the future kitchen, protecting people from drinking unhealthy water. This attractive future perspective is especially sensitive for, small children, elderly people or other individuals who are vulnerable to this type of contamination.

As a safety concept, the tap water sensor experience provides an interesting aspect of possible commercial introduction, e.g., the general electronic tongue concept has been specifically fine-tuned towards the requirements of a taste sensor system, which is able to detect small changes over time in the total quality aspect of drinking water. The measurement procedure, when monitoring the quality of the drinking water, uses the same principle (Fig. 6.17) as the previous simplified device, as shown in Fig. 6.20. By using the technique of voltammetry (Fig. 6.18), pulse cycles of different amplitude and an advance signal handling process, the sensitivity can be increased and the discrimination between samples will substantially be improved.

Depending on the chemical content in the drinking water samples in conjunction with a very sensitive method, it is always valuable to be aware of the tendencies of variation that may occur and explain it to natural phenomenon or failure to measure correctly. However, the sensing focus in this specific electronic tongue concept, as shown below in Fig. 6.21, is typically based on a number of specifically tested working electrodes, however, in this case only from gold and a platinum material. This experience is due to acceptable response sensitivity and selectivity received on drinking water experiments, and its behaviours when contaminated by typically and expected chemicals or biological compounds. In connection to the working electrodes, a reference electrode of stainless steel is, in these experiments, needed to achieve a proper performance.

Figure 6.21. The electronic tongue prototype device mounted on the water tap. Photo courtesy and copyright Peter Wide © 2010.

After receiving the sensor data, a complex data handling and organising, followed by data analysis operations are performed and the outcome, i.e., a measured value of the water quality, is presented. The communication to the user in this specific application is related to a traffic light concept, a green light indicating that the water quality has not changed and a red light indicating a substantial change in water quality has occurred. If needed a yellow light may indicate that a small change has been detected, but in principle not in alarming amounts. In this context, the measuring principle regarding the water quality can be seen as an electronic tongue detecting, identifying and alarming exhibit an external structure of artificial tasting device with visual output properties. From the user's point of view, the external tasting information is presented visually, by different colours, that indicate three possible levels of drinking water quality.

The operational principles differ widely between the gustatory and auditory measurement techniques. The design of an electronic nose concept is quite simple, when using commercial single sensor types that are carefully selected by its selectivity to the measurable parameters we actually want to explore. By combining these single sensors in an array of other similar but complementary sensing capabilities, the achievement experiences a complex system of locally gathered sensors and constitutes as one overall sensor system device, however, with a coveted multidimensional capability. The concept is usually built upon a loosely defined parameter, as for example aiming to provide qualitatively measured values of the air quality, which typically requires perceiving from many sensor elements to capture the complex qualitative characteristics of a "true", or at least expected, definition of air quality. This approach also requires an analysing strategy, that have to perform an experimental design process, to be able to define the concept

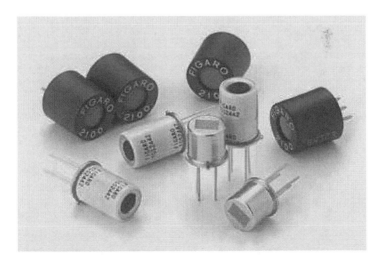

Figure 6.22. A single gas sensor from Figaro Engineering Company. Image courtesy of Figaro Engineering Inc. © Figaro Engineering Inc. 2010. All rights reserved.

of air quality and then relate defined reference data to the device, e.g., to make proper reference tests.

One of the major manufacturers that has been providing the market with advanced MOS-based gas sensors, Göpel (2001), is Figaro Engineering Company, whose sensor models have been commercially available since the year 1968 (Fig. 6.22), Figaro (2009). The original sensor operational performance provided a semiconductor device, which were able to detect very low concentrations of combustible and reducing gases. Further, the sensor element was introduced to operate in conjunction with a simple electrical circuit. The Taguchi gas sensors today, are still easy to operate and the wide combination of different odour sensitivities available is accessible when designing experimental sensor arrays as seen for example in Figs. 6.6 and 6.12. The selection of suitable gas sensors, maybe complemented with other types of sensors, will provide the specific design of a device. Different options in operating with a complementary and specificity that will result in qualitative value response is illustrated for example in Figs. 6.11 and 6.13. The following references exemplify new introductions of possible application areas, that at the time of publication were inspiring to further develop the field of artificial sensor systems, are given by Feldhoff (2000) (fragrance), Bourgeois (2003) (wastewater) and Wide (1997) (air-quality).

The really challenging development phase is considered to be a combination of different systems, and to integrate the capabilities into an overall system performance. A mobile robot and an electronic olfaction system (as previously seen for example in Fig. 6.6) that in combination will provide an autonomous unit with extraordinary capability may exemplify the integrated system function. An autonomous robot detecting leakage of airborne compounds can be illustrated in Fig. 6.23, and provide a mean of identifying the location, i.e., the source of the contamination in the air, Loutfi (2006), or simply act as a companion service robot, Broxvall (2006). The electronic nose device presented in this application coordinated with a vision system that jointly controls the robot's actions. The system first identifies possible objects that can be connected to an odour, or more likely to contain the leaking contamination in the room. The strategy is simple, when identifying suspicious objects, the robot moves closer to make odour detection. The situational awareness in the experiment is more advanced in a real situation but the principles are still similar.

This type of experiments, as illustrated in Fig. 6.23, have of course a wider future perspective to act in a context demonstrating more complex achievements. The need for developing devices that are able to patrol a wide space in the environment for an extended period of time is obvious. The capacity is then to explore the surroundings in order to indicate environmental changes and act as an autonomous early warning system. An additional value is that these autonomous systems can be used for inspection in hazardous environments. These circumstances may arise in areas, e.g., nuclear plants, where people have to work with an increased risk.

Figure 6.23. The illustration of an electronic nose guided autonomous robot. The scene is viewed from another robot's visual system. Reprinted with permission from Loutfi (2006).

As a final remark of the type of application presented in this section, there is a motivation to highlight the performance, when conducting the experiments described above. The outcome of the operational principles has the aim to focus on at least one of the main properties below. The design of measuring qualitative parameters related to artificial- and perceptual-based sensor systems as described in this book may have the following concept basis:

— characterisation,
— discrimination,
— change detection.

Characterisation describes the performance, where the observed samples are clustered by similarity and have been essentially performed with superficial and test samples.

Discrimination performance has been performed to build-up a suitable set of data by repeating such measurements over a relatively long period. The aim is to check the coherence of the system and the ability to discriminate between different types of units, for example, different brands of bottled water.

Change detection capabilities have mainly been regarded in situations where a request to relate the upcoming result with the normal mode, for example, an electronic tongue that detects unique contamination in a occurrence of normally clean drinking water.

The next chapter will comprise the concept of artificial perceptual systems with a focus towards the emerging interest of perceptual prosthesis as a fundamental property to gain artificial human-based qualities. Maybe future prosthesis will also provide the user with extraordinary sensing capabilities.

Chapter Seven

The Artificial Perceptual System — In a Perceptual Prosthesis?

We define a sensor information system, based on the perceptual communication between a human and a perceptual system as:
A perceptual system that produces information, which comprises an array of sensors that may compose of different sensor types, with partial and overlapping specificity/selectivity and appropriately applied mathematical pattern recognition procedures. Also an artificial perceptual system will provide qualitative output values in a human-system symbiosis, Cooley (2008), that can with advantage act as a complementary part in a perceptual based-human prosthesis.

The above definition clearly defines artificial human-based sensor systems as a "natural" interface attached to the human body. However, artificial human sensors that complement the human perception are of certain interest, since it may increase the overall performance in exploring and acting in our environment. Since the human perception is built upon the basic need to survive when exploring and acting in often dangerous and unstructured environments. We certainly could take advantage of a system that will increase our capability.

The ancient behaviour of feeling unsafe forces us to take control of our proximity, by perceptual sensations. The perceptual abilities and skills are related to basic survival functions, principally connected to danger and food. The ability to fight for survival depends on a variety of structural situations, such as if we are able to detect an enemy at a distance, will hopefully enable us to put ourselves into a more secure place. Also, the perceptual capacity helps us to detect and discriminate edible food. This survival ability has during many generations been strengthened by refined abilities in our perceptual system, resulting in the fact that we managed quite well — *since we actually survived*. However, in the last two to three centuries something has happened with our sensing capabilities and a change in the basic perceptual inherited structure is indicated as fading out.

7.1 INTRODUCTION

The measuring operational principle of a perceptual-based sensor composed of different sensor type is characterised by an individuality that makes the selectivity unique when the different sensors are individually combined, as described in an earlier section. The sensor principle has to consider the complexity of each individual sensor's individual performance, when combined to a joint output, because it has to take into consideration all the various aspects of sensing fuzziness of each sensor. In most cases, there is a lack of well-defined parameters, e.g., the value of water-quality or air-quality, when comparing the definition with an individual user's qualitative definition, when hopefully being in a unstressful mode.

Therefore, one single sensor, or sensor type, indeed cannot normally perceive all aspects of the requested information in a specific quality measurement. Consequently, the multi-aspects considering operational principles as an overall system approach is necessary in order to achieve the requested quality parameters of interest. The quality of a measurement approach will be further discussed in the following section. Typically, input information to an artificial sensor system is often in a multidimensional form, which puts demands on the choice of sensor performance or in case of a sensor array, designed to meet the required multi-dimensional functions. This is an important factor when the sensor system tries to obtain the happenings in the environment and transform it into an output qualitative value corresponding to the sensory input. When handling complex parameters in the environment, the strategy then describes a direction from a single sensor solution towards a solution with a number of sensors, e.g., a sensor array. By assembling arrays of slightly selective sensors, and maybe also complementing with sensors of other operational principles, it will then be possible to strengthen the multi-dimensional approach and detect complex parameters in the defined measurement range.

The definition of a sensor information system in the previous page can be further extended. We define an artificial perceptual sensor principle as:

A perceptual system using slightly selective sensors that produce a multi-dimensional concept of information that is often comprised in an array of sensors with partial and overlapping specificity and connected to an appropriately applied mathematical pattern recognition procedures.

Exacting demands in the measuring of complex human-related qualitative values has the prerequisites to also emphasise on the artificial system's performance.

The present trend in sensor developments is mainly focussed on the ability to handle the ever-increasing information flow that is coveted in the society. By using the fundamental understanding that sensor or sensor(s) have to produce specific and local related data to be delivered in time, also where cost is a competitive factor. There is therefore a need for a more complete insight of merging real time data with earlier knowledge and other available information. Further, in case of achieving requested effectiveness, the interaction should be presented to the individual user at the most appropriate time and in an acceptable quality. In the future, the amount of requested information will be highly focussed and competitive. The most determining factor for a system's credibility is considered to be how the artificial sensor system is able to present the complimented information to the user. Proper planning and effective control of a sensor system that is intended to provide essential and requested information at the time of delivery is an absolute condition in order to achieve a smooth interaction. Therefore, it is of most importance that the interaction performs a real-time integration ability between the information collecting level and the information integrating level in order to exhibit the best possible use of information in the following decision-making phase. In the next section, two aspects of conceptual operational principles are presented, that will increase the individual's capability to handle multi-dimensional information flow. However, both systems present an increased human-related cognitive relationship and may be more suitable to human information processing than normal presentation systems.

7.2 THE PERCEPTUAL-BASED INTERACTION

The interaction between an artificial and perceptual-based system and an individual's perception is a crucial moment in the communication performance. The essential concept of how to interface human perception may be of importance in understanding, and making use of, the concurrent consensus that the human capability of 1000 Mbits/s inflow of information received to our senses. Further, an understanding is also needed, concerning the process of how it is further processed and used for decision-making. According to Norretranders (1998), we perceive through our senses huge amounts of information, however only a few percent is consciously used. In Fig. 7.1 below, the estimated information flow through our different sensing organs is illustrated. It can clearly be noticed that the human vision is the most information-providing organ in the body. Therefore special attention will be given to the human visual interaction with specificity of

THE BANDWITH OF THE SENSING MODALITIES AND AWARENESS

Sense	Sensing input	Awareness bandwith
Vision	10 000 000	40
Hearing	100 000	30
Skin	1 000 000	5
Taste	1 000	1
Smell	100 000	1

Figure 7.1. The perceptual information flow [bit/s] through our different sensing organs and the respective awareness [bit/s].

communication based on the system operator that attracts human preferences. In the following sections conceptually defined approaches are exemplified in industrial plants.

7.2.1 The Visual Communication Concepts

In a number of existing complex industrial processes which produce a normal sequence of non-linear behaviors, there is a desirable need for clear and user-friendly communication. It is of importance in order to enable the individual operator as part of the process, in order to perform complex decisions. It is crucial to be conscious of and have an exceptional knowledge about the momentary process states, rather than some generally occurring and specific known physical, biological or chemical properties. However, these basic and quantitative values are often used and the predefined quantities. Pressure, speed or temperature, are very often related as controllable properties and are moreover often shown as the main communicating parameters to monitor a process, e.g., in a paper plant or food processing line.

Normally, the industrial process makes use of different presentation models to visualise a complex industrial process. Then, depending on the presentation mode, the operator's involvement and skill differs in more or less advance interactions as a variable part in the process control loop.

Typically, in the industrial process communication with an operator, the visualisation modes normally are implemented as:

— *a physical view of the process or parts thereof. This can be performed by virtual reality.*
— *an implementation structure mode uses logical or physical block description.*
— *a mathematical mode that visualises the process or parts thereof by a mathematical description. This mode requires an understanding from the operator that enables the user to be able to adjust the relationships among the parameters in the process.*

Both the illustrative examples (in section 7.2.2 and 7.2.3) below use the mathematical mode to effectively introduce and basically involve the operator into the process. The operator in conjunction with the presentation technique may then in the mathematical mode act more effectively in adopting the communication aspect than the other of the above modes. This can be indicated by the use of single features in the interaction, designated as human-based featural singletons in both examples.

Traditionally, visual presentation or interfacing systems have prefered the type of complex, colorful and computer oriented presentations. Over the last decade however, it has become increasingly obvious that a more effective relationship between the human operator, conventional information and interacting systems. The effectiveness is needed to secure the communication built upon the capabilities and abilities between the two parts in symbiosis (human and system).

Since the human operation is still often used for complex industrial control and in decision-making situations, an interface that is convenient, useful and effective from both human and technical point of view is not only required but also crucial. The need, therefore in many applications, where product quality, economic efficiency or security is focused upon a rational and correct handling of the industrial process are desirable parameters to be achieved.

The real time and momentarily performed action of an individual operator is however dependent on experience gained earlier in coherence with unique and detailed knowledge of the expected systematic process behaviour. The specific operator power is particularly valid in abnormal and unexpected situations. The visual and symbolic interface system proposed below, however, show examples of a new and different approach for communication, the visualisation of the actual and total process state which includes an ability for prediction, if necessary. This concept is also a way of securing the identification of different states in the process, by using a skilled operator who needs to be more of a generalist than mathematical or control expert. The illustrated system will provide a natural indication with colours, position and shapes to view the overall process status. By following the natural human prerequisites, the operator is able to change focus from the individual process parameters to more overall, complex and human adopted interface.

The two following examples intend to show the ability of presenting additional sensor (and other) information to an individual (operator) in a manner, by comprising an integrated and effective information flow with an adaptation that optimises the communication between human/computer based system capability. The visualised process interface then secures an effective transfer of intercommunication and guarantees an adjustment to the weakest part in the interface, which in many cases may be the human operator. Therefore, there is an obvious need to ensure that the information will be presented to the operator in a trustful and intelligible way, so that proper decisions could constantly be made. Thus, we need to know, within a reasonable reliability, that the operator will not discard the attention, i.e., the system has to ensure that attention has been established. Instead the individual needs to pay attention to the information presented and consciously add the information to the ongoing process before making suitable decisions and interact with the process. Even if the information is of no or minor use at that specific moment, it has to be validated together with other information. Only at that moment, the human can be acting as a part in the process loop.

7.2.1.1 Visual Communication Interface

In order to let the operator be more involved in the real time industrial process structure, without degrading the operator's ability to follow and influence the process, then the effectiveness in a transparent but user-friendly perceptual interface is of high importance. In the following approaches the operator is able to communicate with a sensor system in real time and interact in a complex process that typically provides multi-dimensional information. The operator, with limited sensing capabilities, needs to be adapted to an interface offering the best information flow and ways of presenting the right information. Also the importance of the needed system to be adjusted to the human limitations in speed and shaping of human-related symbols, like colours and shapes, will increase the interaction. The following examples shows that in complex processes, there is an obvious need to change the focussing on feature singleton and quantitative values, describing only indirect relations to the overall process condition. The two approaches below will show examples of complex visual communication systems indicating the process behaviour in conjunction with human capability.

The approaches shown in these two applications that are based on two different techniques, use advanced technology, based on control theory and chaos algorithms respectively. The common issue is however the human-based qualities that may effectively attract the operator's attention, i.e., visual attention-based indication.

7.2.1.2 Operator Portal

In the illustrated example, as described in Section 7.2.3, the glyph representation concept is represented as a glyph graph, in order to visualise large data sets. It is then possible by using a glyph graph to explore the variable relation to each other. In the industry, as in many complex related areas, there is an obvious need to transfer information from advanced and ill-structured processes into more compressed and secure information flow to ensure the overall performance. The main issue however, is the procedure of how to present essential process status to an operator, who normally does not exhibit the skills of a mathematical or control specialist, to handle a huge amount of information in real time. There is a therefore a need to present a reduced dimensionality in interfacing from an n-dimensional information plot into a two-dimensional information chart. The presentation portal is actually the operator room, where the operator is able to detect activities in different process states in real time as well as receiving an overall process survey. The process presentation system has a visual impact, since it facilitates the human-process interaction. The visual context is demonstrated by underlying algorithms for control purposes and provides an effective human-system interaction.

The operator portal shown in the Section 7.2.2, is another type of visual communication. The chaos-based concept approach will attract human-based singleton. An overall visual presentation is presented that indicates whether the process is within one of totally three existing modes:

— within a normal mode
— outside the limitations or
— on the way to become uncontrolled.

This example will provide a controlled state of presenting information in a comprehensive way. The presentation approach in the chaos-based concept communicates the process's overall status with only one single presentation event plot. On the other hand, the glyph graph also examplifies an additional feature of detailed information in a number of underlying presentation plots. However, both examples show the intention to make the operator more integrated and involved in the process behaviours by visual interaction and extended possibilities to provide proper (control) actions. This industrial-directed communication can also be seen as a degree of symbiosis between a process and an operator.

7.2.2 The Chaos Concept

The artificial system's communication with an individual person can be massive, using all available sensing, and may comprise earlier experiences as well as obtained knowledge. The human is highly complex but also very vulnerable to other internal conditions, that is a system with stress, and external conditions, e.g., disturbances. If an individual does not get enough sleep, food or exhibits stress, unpleasant temperatures or is in a harsh environment, then her judgment will change dramatically. These disturbances also affect the complete perceptual system, and as a consequence, the individual ability to perform well as an essential part of a process control is limited when sensing, making decisions and acting together with a system.

The interaction between different actors is crucial when considering both the human and system as an integrated unit in a complex and symbiotic process. There is indeed a need for the communication to be in such a way that both timing and information can be processed effectively by each part in the system. The interface consisting of an artificial human-based technology which provides the condition that is effective for both the system and operator ability, is considered a necessity when applying to intelligence in a symbiotic relationship between operator and artificial system. Since the human capability in some situations can be considered to be lagging behind other artificial parts in a system, it is essential to consider the communication preferences and, in this case, the visual presentation interface, in order to minimise the drawbacks of the involved individual actions.

There are many reasons to believe that we usually visualize the real world reasonably accurately. But we have also reason to believe that the retinal image of the environment will not give a complete and correct picture of the world (whatever now is considered to be the right picture). However, the combination of some basic distinct visual cues that convey information about the environment, with the perceptual experience achieved may provide results in a construction of a more optimal visual representation described as a visual presentation interface.

The eye, which acts as an initial sensory receptor, provides us with an image of the world around us. Initially, the vision sense converts light energy into a pattern of activated neurons, where each active neuron represents a brightness value and a spatial location. Hence the neurons, or group of neurons, encode the visual information. The process is parallel and acts over the entire visual field and is used for encoding. The preconscious mode is followed by a subprocess that makes combinations, in order to construct a unified representation, in the visuo-spatial code. The visuo-spatial code is composed of three sub-codes obtained by encoding separate representations of the input, Glass (1986).

These representations are:

— brightness/colour, (including black, white and shades of gray)
— shape, and
— location.

Finally, the process of the combined representation of the input vision sense is compared with existing representations in the memory, Wide (1996a).

The information from the sub-codes must be combined in order to form an integrated visuo-spatial representation of the input vision sense. A representation of an object can be compared with a memory representation before the definitive integrated perceptual representation is formed. This is sometimes called selective attention.

An object, in this illustration, exemplified by a communication and monitoring application that corresponds to a symbolically displayed visual plot and may be described as a distractor on some salient dimension. This can be referred to as a featural singleton, which is expected to differ from all other stimuli. It can be shown, Best (1993), that singletons can capture human attention, but only when they are some how relevant to the perceiver's goals, for example, when the individual knows that the symbolically displayed visual plot makes it possible in defining a behaviour. In this case of behaviour, then the specific featural singleton is processed with a high priority (salient).

Under normal conditions the image shown on the visual presentation system in Fig. 7.3, works as one among many lasting impressions, as a communication input to the individual's visual perception. This dynamical process normally undergoes a constant flow of course of events. When motion occurs in the visual range, the individual's visual perception has to consider that momentarily we focus on moving features of the image on behalf of other behaviours in the same part of the environment. There is significant evidence that a specialised neurophysiological part of the brain exists for motion processing, which uses an extremely rapid and complex computational capacity, Stillings (1995). The visual motion perception is apparently able to track identifiable visual elements such as edges or lines over long time spans and distances. This human visual capability has a remarkable ability to process and utilise motion information and may be one effective communication interface when fast and reliable information is needed.

The visual presentation approach was designed to exemplify an effective human-system interaction and increase the overall systems performance in industrial presentation systems. The presentation is shown as an application solution using a symbolically displayed visual system and has been tested in the process industry with satisfactory results. The illustration has been built upon the dynamical system according to Fig. 7.2 and will present different process status dependent on the multi-sensor inputs. The sensor model, described in Fig. 7.2, illustrates the dynamical system, based on charging (source) or discharging (drain) a capacitor (store) controlling a by switch (gate) with a sensing unit. The system dynamics is affected by the sensor signals and represented in states by arranging from stable

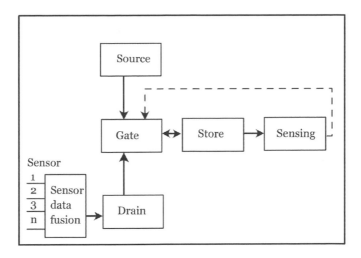

Figure 7.2. The presentation of the process status as shown in a chaos based graph.

through periodic to chaotic phases, which is monitored in a visual plot, as shown in Fig 7.3.

The proposed illustration, seems to be if not an optimal solution, then considered to be a highly effective communication, by using the human preferences in the preconscious coding properties, when designing a display and monitoring system that attracts the individual's sensation-based conditions, Wide (1998a).

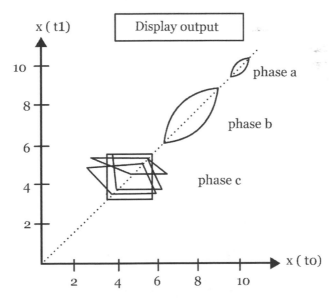

Figure 7.3. The visual interaction in a chaos-based attracting visual perceptual system.

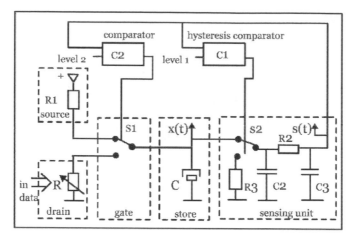

Figure 7.4. A simple electronic circuit is needed to create the different chaos-based modes.

Colour, shape and motion are considered as important human visual features used in Fig. 7.3 and may be the most effective attentions directing processes that attract the human process of attention. As an example, chaos systems may frequently be used in the future, because it is an excellent feature that attracts the power to the accessibility of an individual's attention. A chaotic behaviour on the presentation graph will unconditionally and immediately attract the visual sight of an operator as illustrated as a presentation image in Fig. 7.3. The technology-based methodology can easily be designed by the use of a simple electronic circuit as shown in Fig. 7.4 above. Further details regarding the system's dynamical model is described in Wide (1996a).

7.2.3 *The Glyph Representation Concept*

The main problem in a complex industrial process is to present to an operator the overall status of the process in real time. The often huge amount of sensor information available may not be a realistic option for the operator to effectively adopt and make use of in decision-making. Therefore, a presentation system that makes it possible to transfer multi-dimensional process supervision into a feasible three-dimensional visualisation solution can be considered to be an attractive interface to the operator. The technique was originally designed as an industry-related application, a human operator interface to a cold rolling of metal process. The visualisation concept for the graphical representation used in this application is a modification of the glyph representation, Törngren (2004).

A glyph graph is often used to visualise large data sets of discrete multivariate data. It is then possible to explore the variable relationship to each other. A glyph is a single graphical element that represents a sensor output and data belonging to the same attributes, e.g., colour, shape, and location, which is mapped to graphical attributes of each glyph. Also other attributes may be of interest to present from a process, like for example an object's motion, surface texture and orientation.

The goal of the information provided is to continuously communicate, by illustrating each sensor/control system in a sub-process behaviour and the impact of their relationship and influence to the total process. This approach allows the operator to be able to adjust, correlate or modify each sub-process in order to get the most optimised process performance. The operator may immediately after each change on the sub-process level also directly be able to observe the overall process response. This exemplifies a system that makes use of the operator in the control loop as a natural interaction and illustrates a degree of symbiosis between human and system.

The communication of a process status has in this case been modified to adapt with the operator's ability to act in the loop of interaction. The given mathematical mode implies that all sub-processes are represented in one presentation, as well as including the presentation of the complete process behaviour. This strengthens the cognitive view, indicating the relationship to separate sub-processes and also the effect of the complete process when adjusting one single as well as may parameter.

The glyph presentation in this example is built upon the view of a cylindrical co-ordinate system, where information about each separate sub-process, as well as the whole process space can be visualised in one single virtual representation. The glyph representation of the whole process space as a transparent ellipsoid is the origin of the cylindrical co-ordinate system and the glyphs representing each sub-process as transparent spheres are placed with equal angles around the origin,

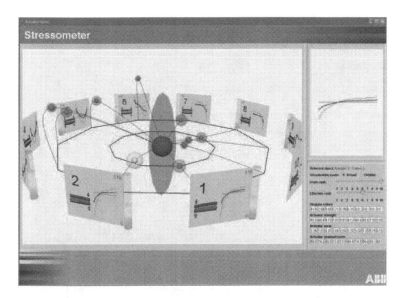

Figure 7.5. The visualisation process of the super-glyphs in a stressometer for worked rolls is shown. Photo courtesy and copyright C. Törngren and G. Fodor © 2010.

as illustrated in Fig. 7.5.

The goal of illustrating the two exemplified presentation systems is to support, and in some sense complement an operator's perceptual abilities, which may not be updated to the underlying complex mathematical control mechanisms. By that means, the underlying concept is not only to understand the behaviour of the process, but also be able to modify complex algorithms and observing the result at an overall level. The presentation system will naturally complement the operator's ability to handle a complex industrial process and to supervise a multi-dimensional sensor control system by the use of effective subprocesses and highly interactive presentation systems. This system is used as an operator interface for a Stressometer system at ABB Automation System AB, Mikkelsen (2004).

7.3 APPLICATIONS

The society of today desires to be more and more effective and highly productive. The society is also involved in an ever increasing technology-dependent interacting process that gives rise to a certain behaviour at work, as well as at our free time. Other social patterns are created by the use of technology dependent devices, e.g., the possibility of carrying out a walking exercise in a gym instead of experiencing the parks, forest or on the beaches as shown in Fig. 7.6. These trends are constantly changing our lives and social relationships, which then require a specific perceptual competence that appreciates the perceptual impression of a gym more than a walk in the nature. The human may then in some aspect be regarded as

Figure 7.6. A spinning round in a closed room with restricted perceptual impressions — as well as a ventilation system.

technology-driven, when substituting natural impressions by corresponding artificial ones. The use of artificial and technology competent devices that complement the human perceptual capability and substitute natural impressions is now steadily increasing. There seems to be a priority to appraise the social networks before the interaction with nature. The development of social-based artificial devices has, since some years now, been experiencing an emerging interest. The need for social interaction seems to be of high importance to human's well-being.

However, the development has often been technology-driven by social trends that lack a social affiliation, where the young generation instead (or nowadays even elderly people) are playing computer games or using a social robot instead of a traditional social interaction in conjunction with the families or friends.

By definition, the social connection to technology-based devices, i.e., automated and maybe autonomous driven sensor systems designed as an acting perceptual prosthesis, which have a performance implying that we are comfortable communicating with the device that provides us with useful information. The requirements of a social device may be directing the following abilities:

— *interacting* with the individual in her close proximity to communicate some sort of perceptual information flow needed to make the proper underlying decision or filling out needed data.
— *understanding* the basic concept of the parts involved, their aims, prerequisite and limitations in the manners that fulfil a perceptual requirement or needs.
— *complementary* perceptual behaviours between the technology devices and an individual's perceptual ability, the prosthesis may increase the performance on the basis level of quality.

The social dimension has to exhibit a certain minimum capability of complementing and understanding the human perceptual needs in interactions with other humans. Since we are able to judge situations reasonably well, get a general view, find the context and use our intuition in common situations, a feasible perceptual prosthesis has to consider the human-related behavioural properties. On the other hand, humans may have a need to complement for the poor capabilities, like managing a number of situations simultaneously, performing tasks by routine, remembering details, abandoning incorrect conclusions or handling stress. Well-balanced information that is presented to the individual at the right times is crucial for the effectiveness in a sensing–acting process, e.g., a perceptual-based prosthesis.

This approach implies that technology-based socially directed devices can be used in a different number of personal applications. In the social aspect, two related applications will be further discussed. The area of *personal interior systems* or more specifically; the perceptual acting and natural mimicing prosthesis, is foreseen to be an emerging field of technology development. Further, the area of *personal exterior systems*; is defined as a remote system or a prosthesis acting like a technical system connected to the human perception. The system may be exemplified by a social network of both artificial and biological agents in symbiosis, can be illustrated in the area of the search and rescue. A socially-related artificial and

Figure 7.7. The personal interior/exterior perceptual prosthesis system.

complementary sensor system, is focussed on the relationship in communicating and delivering the required and proper sensing abilities and effective control of the human-related environment. The function has in the field of medical technology an important feature to reconstruct new functions in replacing missing body parts. In some way the given systems can be seen as prolonging the human sensing abilities, if concluded in a perceptual acting prosthesis. In a wide perspective, the artificial prosthesis device may exhibit functions that are more advanced than a corresponding body function for example in symbiosis with human as shown in Fig. 7.10.

7.3.1 Personal Interior (Prosthesis) Systems

The human body is today a functional unit that has been developed and refined during generations. Its parts smoothly perform the functions of fine-tuned, integrated and coordinated entirety. This complex functions then creates a complexity and in some sense an unique human. However, due to coincidences in life and certain unforcseen occations which may happen, e.g., by disease or accident, which can result in change in our performance. This may result in loss in abilities and variations in functions between individuals and the need for replacement of functions may then be obvious. Nowadays, limbs can be replaced by artificial solutions. The inner organs, e.g., kidney, heart, etc., may be replaced and the medical competence today is able to make the human to integrate with high technology-based functionality incorporated as spare parts in the body. The next development phase may include the direction to make perceptual prosthesis available to human prosthesis selection.

The skin, however, may already today be substituted by artificial-based developed skin. Tactile hand prosthesis has been presented in the literature and the interface between the neural pathways and a hand is now available. Today there is a possibility to make micro-solutions when designing an electronic nose system and integrated into a pair of artificial visual-based spectacles with, for example, vibrating outputs in connection to the skin. The same premises already exist for the auditory prosthesis to be placed in or in connection to the ear. The development advances in finding, tracking, and identifying objects in an artificial vision system, e.g., a camera system, has to make decisions on very large distances with high performance. These premises will hopefully find a market and be generally available to strengthen the development of artificial human perceptual systems.

The ambition to replace or complement the procedure to fill the mouth with an amount of liquid or food mixed with saliva, to provide proper sensing to decide the gustatory properties, is maybe hard to realise as a traditional prosthesis device. However, the pictures in Section 6.3.2, as well as Fig. 7.8 above is illustrative and illustrate the principal solutions. The proposed sensing devices are placed at a distance to the body and may be an effective measurement procedure in conjunction to the drinking of water or testing food. These systems are then considered to be a personal exterior sensing device. A prototype of a possible safety device for drinking water is shown in Fig. 7.8. In this application, the sensor is mounted on tap water in kitchens and used by the whole family.

On the other hand if the perceptual gustatory sensing device is placed in a very close conjunction to the mouth, however still outside the mouth, could this solution be classified as a personal interior perceptual prosthesis? The distinction can be further discussed, but the definition made in this section is that a personal interior perceptual system is somehow in direct contact with the body, i.e. the very close proximity of the human.

The personal interior perceptual system is also required to be considered as a natural part of the body as illustrated by the person's left hand and arm in Fig. 7.7.

Figure 7.8. The taste sensor device for drinking water mounted on a water tap in the kitchen. Photo courtesy and copyright Peter Wide © 2010.

From a social related perspective, the interior prosthesis should be adopted to the body, providing a natural appearance and human-like perception abilities as for example compared to the system device shown in Fig. 7.9.

The gustatory sense is specifically vulnerable since a safe solution must be considered to be able, in real time, to measure and in attaching the object outside the body, i.e., in this case outside the mouth cavity. There may be an attractive solution, where the gustatory device is placed in a feasible solution by constituting a cavity of sensing ability and a first response appears before the object, e.g., the liquid, enters the mouth. A prototype device can, with advantage be implemented in an ordinary drinking glass device, as shown in Fig. 7.9. The prototype has been realised as the functionality of a perceptual prosthesis device and its performance is verified as an effective complement to the human gustatory perceptual sensation.

The technology improvements will exhibit an ongoing procedure where new and emerging solutions are foreseen to be pushed towards the individual's needs

Figure 7.9. The perceptual prosthesis device, complementing the gustatory ability. Photo courtesy and copyright Peter Wide © 2010.

Figure 7.10. A humanoid robot mimicking the human performance, with human-related abilities. Photo courtesy and copyright Peter Wide © 2010.

to sense and act, that will result in more or less a constantly increased performance. The artificial prosthesis parts are foreseen to be frequently used in the human body and move in a direction toward adopting further functions to the individual's specific behaviour. This raises an important question about where do we put a limit in defining a human as still being a human and when does the human go over the threshold to become an artificial agent, i.e., humanoid as for example shown in Fig. 7.10. Today there seems to be a broad understanding in defining the distinction between a human and an artificial humanoid, but when we start to mix the parts from each other, then we will probably end up in a fuzzy challenge when defining a human being. Hopefully the time for this type of discussions is not here yet, but we can be sure that the time will come, in the year 2020, 2030 or perhaps 2050.

7.3.2 Personal Exterior Systems

Complementary systems for increasing the human perceptual ability are of an emerging interest that are capable of performing advanced interaction with a human operator. The attractive possibilities that can be achieved when investigating conditions at a distance from the human are undoubtedly a flexible and safe solution. The performance, without interfering in an unhealthy or dangerous area of interest, for example, due to personal risks, is of importance in the area of personal safety and security. The human safety and security aspects may with advantage use supportive devices that are able to provide complementary performed abilities, and based on a distance evaluation from the occurrence.

A typical application in the peronal exterior sensor system approach can with advantage be demonstrated in the search and rescue applications. These types of applications are more frequently used in remote operations as a joint complement to human perception. The search and rescue operations are often built upon different skills in heterogeneous groups acting and interacting between humans, dogs and robots. The concept of interactive technology systems, where both static and mobile sensors are used to enter a hazardous and unknown area, e.g., a burning building, an earthquake or other types of disaster situations, and in close interaction with a human operator can be illustrated in Fig. 7.11. The human participation can be seen as an active part in the communication loop that performs the strategy and decisions, i.e., the artificial perceptual sensor systems may act as an essential remote function to perceive information in the hazardous areas of interest. On the other hand, the sensor systems can act as an autonomous object, providing the human operator with requested or other relevant information.

The technical support systems that complement the search and rescue personnel in an environment, that is known as stressful, intensive and hazardous, are aimed to find relevant information as soon as possible. It is considered to be a trying, time demanding and often challenging task for the heterogeneous response teams to smoothly interact with the environment with the aim to effectively perform results. The main issue is normally to perceive essential information and to manage these data in real time. This can be considered to be an ultimate ap-

Figure 7.11. A humanoid robot mimicking the human performance at a distance from the human operator, cooperating with human-related abilities. Image courtesy of QinetiQ. © QinetiQ Ltd 2010. All rights reserved.

plication for sensor systems in supporting humans and extract the human sensing capability. These operator-driven exterior sensor devices complemented with additional external remote information sources, may even be acting and integrating with mobile robot platforms performed in the search area.

The human-based sensor system is aimed to remotely interact with the environment in order to achieve the best possible information flow from a distant disaster area to the rescue workers. Then, based on the data received, it may perform substantially safer search actions, and make more effective decisions than without the complementary artificial information sources, e.g., Das (2003).

Consider a purely hypothetical case of an accident, when suddenly a disaster situation occuring in a search and rescue mission, in which a heterogeneous response team (humans, dogs, robots — movable and static sensors) enters an unknown building. The different agents, via a network are connected to each other and will each receive unique time-dependent information. Since different agents have different specific tasks and conditions, then there is a need to provide each agent with specific real-time amount of essential information. The communication between the agents is necessary for tasking the network, sharing information, and for the overall situational management. The available performance of each agent will be optimised in the sense that a camera will provide and share visual information and an olfaction device distributing smell properties to the team. The management, e.g., the team leader will then have access to all information and based on the person's skill and earlier knowledge be able to extract essential information for decision-making purpose. Also, the tactical methodology has been agreed upon. When they arrive at the scene, the tactical operation starts, and a general strategic structure is considered to be valid. The strategic path is continuously updated and refined as the procedure goes on. The advanced information needed by the manager and the valid information provided to the different agents is of most importance both in time and amount of data. In fact if human agents are involved, then every agent in the team — artificial, human or dogs — on the scene should be following the same structural level of semantic and organisation when interacting in the network. The danger assessment strategy is estimated and the different agents' performance will be validated. If there is a risk for humans entering the disaster area, the artificial mobile and static sensors specific benefits can be used. Then the different agents, or sensing devices, can start producing data that, e.g., may be used to establish a generic map of the scene. The map is a knowledge estimation of gained information that is continuously updated and agents of a characteristic static behaviour are placed in strategic places. The communication is frequent and the agents are updated with essential activities about, for example, injured people and events that may affect the rescue and search strategy. This procedure continues until the coarse scene has been searched, and new tasks will be followed. The hierarchy in the network may already be settled when humans strategically manage the actions, the dogs are extremely useful to find people in the debris but their communication may be limited at a distance and in a chaotic environment. However, artificial sensor systems may be the solution in contaminated areas, where the risk for people and dogs are impending. The logistics seems simple but the interaction between a number of actors in a chaotic environment is indeed a complex task, Murphy (2008). Instead of risking human lives at dangerous scenes, the network of specialist actors comprising of the latest technology capabilities will complement the human involvement by perceiving the sensing information by intelligent artificial perceptual systems. This is however a challenging task, where the overall solutions have not yet been found in the literature. Improvements of specific agents in the network are however, an area of research and technology advances can be seen relatively frequently, e.g., inorganised robot search and rescue competitions.

The ultimate goal in such an intelligent network composed of different specific skills, is when each object can be regarded as a unit with human-like skills, performance and communication abilities. This means that each agent behaves and relates in a similar manner to a human, and no difference in performance may be discovered from the communication in the network. As mentioned earlier in Parlin (2007) the movable perceptual and sensing agent may with advantage be acting as an involved individual, observing and reporting from the hazardous environment. This is considered to be a vital part of the network's intelligent performance. The team manager will, at this moment, make decisions based on the remote information gained from the different available sensor systems, i.e., agents, in an information flow network. The exterior system acts as a perceptual sensor system providing essential data to someone collecting, sorting and analysing and prioritises the information. The information will then be integrated with other sources and merged together into proper decisions, e.g., chemical sensing controlled by mobile robots, Lilienthal (2006).

The first responding team will indeed have the need for technology improvements in order to complement the picture of the situation and evaluating other essential information. This methodology of a remote sensing concept certainly improves the performance of the information flow extracted from received information from specific agents' ability to present the specific data.

Especially, the remote perceptual sensor system approach would have a focus on:

- *localisation* and search path in an environment with no or minor infrastructure such as a collapsed building, as provided to the involved agents.
- *agents' information flow* across an *ad-hoc* network that are able to make proper localisation on the basis given, for delivering the most relevant and current information in time to other agents and to update the pictorial map.
- *using global information* available in the network to provide an extensive source of information base.
- *interacting with management* (human or artificial system) in the information system and gathered between the active agents involved.
- and last but indeed not least, the *information logistics* is of most importance. Since the use of multi-agent structure involving humans with special skills, animals

and artificial technology-based specialist systems have to work in a similar environment and require advanced communication between them, the performance shapes an unified team that has to work on an equal basis.

The fictitious scenario described is in its totality and dimension not technically available (at least not yet seen in the literature) as proven solutions to achieve the required intelligence. This fact is mainly due to the limited performance of the single involved agents, and of course also the lack of intelligent network interaction between the specific parts. However, there are single social robots exhibiting more or less autonomous behaviours, designed for complementing the individual's lack of sensing and communicating information, i.e., solitude of lonely people. These artificial agents have up to now mainly focussed on basic abilities, for example partner robots, Toyota (2009), service robot, Saffiotti (2008), and specifically towards the elderly and disabled, Harmo (2005).

The ultimate goal of such an action-driven network is that every agent is an active part of the team that provides the best possible jointly handled information. By those means, we may achieve participation in a loosely coupled information network that have a minor specific interest. A human individual, artificial stationary, mobile sensor device (or even a dog) can i general terms be networking, as long as proper information is delivered at the right time following a joint strategy.

The application presented in this section is meant to illustrate the methodology of a realistic remote perceptual sensor system. However, similar principles can be seen in global warning systems for tsunami, earthquakes, etc.

There is surely a technology-driven development toward true symbiosis between human and artificial system that cooperates in an effective way. The cooperation can be bi-directional as for example when humans report via mobile phones observations in the environment to an artificial systems, as described earlier in Section 2.12 and 5.5, Paulin (2007) or when autonomous robots are picking litter on the streets and parks, Mazzolai (2008). Then we really make use of the symbiotic effects in cooperating between humans ans systems.

Chapter Eight

Conclusions and Future Works

The idea to merge human and computer capability into a symbiosis of interactions – is indeed not a new idea. With the introduction of the computer in 1950s, an overwhelming belief in the computer and its possibilities arised. The strong conviction was generally accepted that the computer capacity would also possess equally good qualities to the human-based knowledge. The computer and its interaction with human was discussed and futuristic solutions predicted a new glorious era. Already in 1960, the following text was written by, Licklider (1960).

"It seems likely that the contributions of human operators and equipment will blend together so completely in many operations that it will be difficult to separate them neatly in analysis. That would be the case if, in gathering data on which to base a decision, for example, both the man and the computer came up with relevant precedents from experience and if the computer then suggested a course of action that agreed with the man's intuitive judgment".

Since that time, the developments of computer performance has been incredible fast and is now expected to match human brain capacity within 10–20 years. Then a possible conclusion, that the challenge for future technology driven development in the field of artificial human based sensor systems is heavily related to how we are able to meet the requirements of designing new sensor solutions and bring new innovations into this field. The success of producing new innovative sensor systems is also dependent on other factors, for example computational organisation and functionality, costs and availability. However, a qualified conclution is still that:

There seems to be a clear indication that the challenge to successfully provide advanced artificial sensor systems in symbiosis with human suggests that new advanced products will be on the market within the next decade.

The development phase of providing new artificial supportive human-based sensing systems, in conjunction with an acting performance, can with advantage be used in designing new prosthesis device. The capacity of a sensing ability is typically considered as complementary sensor systems relying on an information flow, that may have a positive effect on a person with limited perceptual abilities.

Artificial Human Sensors — Science and Applications by P. Wide
Copyright © 2012 by Pan Stanford Publishing Pte Ltd
www.panstanford.com
978-981-4241-58-8

Then it may be appropriate to also consider an artificial and human-related sensing device, e.g., an artificial tongue, nose, ear, eye, etc., as a sensing prosthesis. The goal has earlier been to spread comprehensive technology trends to a variety of human-related activities within industries, services and in private homes. The symbiosis-based technology, when integrated with the emerging infrastructure for global information, is however foreseen to have a great impact on our daily activities. It is expected that this emerging technology will open business abilities, huge opportunities to truly improve the quality of life and freedom to sense the world on an equal basis whether young or old, healthy or disabled. The significance of usability is foreseen to be of a similar scale than what we currently experience, by for example the introduction of the personal computer or later the unbeleivable development of networking the information flow, the Internet.

As the human perception is instantly improved by artificial means, we will most likely experience more of complexity, where for example artificial devices are able to depict the environment in a more advanced picture Then humans may be able to sense with complex sensor-based biosystem including advanced processing computer programs. At that time when it will be frequent available and we are able to commonly make use of the benefits and possible applications based on human technology. Consequently, additional effects in intelligence and experience will also increase, where complementary sensor systems are expected to evolve into additional symbiotic information flow operators. These systems, or symbiotic sensing prosthesis, will act as individually adapted supportive devices which probably represent the next evolutionary phase of information, comprising of a new intensive sensing and interaction paradigm.

The emerging focus will probably be upon the aspect of user-centred technology, where human demands for complexity, timing and individual adaptation most likely will challenge the requested performance. The emerging technology will then integrate computers, sensors and communication into effective demands of the system functions that are jointly integrated in task-specific networks of systems. The expected trend aims for a high performance information flow in a variety of directions. The future trends will probably also shift from today's technology-driven and computer-based industry into the human-requested development involving a highly advanced technology in harmony with an individually centred and consumer-oriented industry.

The information collection in specific individual requested devices would most likely be able to learn and adapt the behaviour of the symbiotic human. In order to improve the system's working conditions, including other technical equipment as well as other human collaborations, the artificial device may adapt to the individual requirements as well as the environment. It is then reasonable to expect, that under these conditions the flow of information will interact with other individuals' sensor system, who will contribute with human-specific exchange capability in a true joint co-operation. Then we will achieve the goal of freedom to symbiotically find and exchange information that supports the best environmental update between individuals. The system-human interaction may then be seen as a symbiotic partnership, in which each partner will lead in defining the process in some situations and in other cases provide companion ability. The functions related to the leader/companion capabilities of dedicated partners will be decided on the basis of maximising the overall performance of the joint symbiotic team, Petriu (2008).

The primary motivation for the artificial human-based sensing concept, is to explore but also generate an expected progress in a perceptive area of human complementary sensor systems. The concept is foreseen to provide a new and emerging capability in human well-being.

This book will hopefully contribute and view an extended approach that directs the possibility of added perception abilities. The future developments will most likely be in a continuos phase of technical and social progress, in order to direct the focus on the fact that we as humans constantly have a need for a more adequate sensing ability.

Therefore, we may intensify our efforts to establish the following influencing aspects of motivation and arguments:

- a continuously growing elderly population with decreased sensing performance, will jointly request the accessibility for technical solutions, that are available on a commercial basis on the market.
- a need is established for a new concept of sensor abilities to compensate for the natural individual acuity that is affecting the variety of human perception.
- a new era of more powerful computer based systems will create performance that is more comparative with human capacity. This will provide a unique possibility, which at the end will create effective human-system interaction.
- a necessity to use existing technology to compensate for disabled functions in perception due to accidents or illness.

This general concept is indicative and has the overall purpose to provide the specific individual needs of complementary sensing, that surely will provide a truly improved perceptual quality. Further, it may contribute to improve the human communication and interaction, aiming to achieve a valued social solidarity, when allowing to be active and dignified. These are human values that today in many cases would not be possible to achieve.

Although a vision for the coming decade may be to provide individuals with the ability to experience more of the world around us and get an extra flavour of new and artificial-based re-created sensations, that maybe since long time has faded away. The technology push, towards the challenging symbiosis between human–artificial behavioural system approach, will then be encouraging in order to find effective human-related solutions on an individual basis.

However, the concept of improved human perception providing attractive and increased values in the individual's performance will of course also exhibit some problems. Individuals who do not actively make enough use of their senses by simply not strengthening the limits of capable sensation, will in the long run exhibit a decreased performance. The genetic adaptation to the nature is a continuous effect of developing the perception abilities. The need to continuously focus, adapting and coping with new challenges originating in the environment is obvious. Thus, the tendency is predicting toward one direction, that is crystallised from past experience and existing needs, indicating the decrease of sensing performance in the human perceptual ability. This ongoing effect has probably proceeded for generations back in time. This paradox will then consequently indicate that when relying on complementary information in conjunction with the human senses, a continuous slow and degradating effect will most likely occur in the human perceptual sensing capability. The smarter the artificial sensor systems are, the less the individual will have to rely on its own sensing capabilities and strengthen its abilities.

The following examples consider different aspects of using artificial human-based sensors that may be of importance to describe the latency of a supportive artificial system. The first aspect points at the natural degradation consisted in a population. Another aspect, however, is the psychological impact of trusting artificial systems, revealling a lack of strengthening the individual's perceptual ability.

Drivers are often aware of their own signals that indicate fatigue while driving. However, a driver often underestimates the risks involved when she is tired. On the other hand, car drivers may, when not focussed, overestimate their ability to control an arising situation. Therefore, dedicated artificial information to support drivers in these situations seems to be a potential countermeasure to help support the driver's ability to increase her awareness. Such information should aim to complement the driver's own awareness of the risks of driving when fatigued. These systems aims to provide relevant and time related information. This effect may result in a psychological feeling of trust to the system without considering the limitations in the specific performance.

A study in Sweden, Anund (2004), revealed that there were differences between groups of drivers, regarding health, sleeping habits, impression of different fatigue-inducing factors and countermeasure preferences. Young drivers and professional drivers seemed to be the drivers who are more at risk of fatigue-related accidents.

A study in USA by Newman (2004) reported that three times more men than women reported falling asleep while driving.

The need of making drivers realise the safety aspects due to fatigue-related situations have forced a safety concept to be introduced by car manufacturers to increase their efforts in finding effective counter measures.

The active system detects the driver's behaviour and monitors the car's movement in order to assess whether the vehicle is in a normal controlled driving phase, or if the system, i.e., driver and car, behave in an unstructured or uncontrolled manner.

The upcoming trend of these systems seems rather to monitor the driving behaviour instead of the human behaviour, due to an increased reliability in the limited performance of an artificial system. Of course the individual's behaviour is more complex to monitor, mainly due to a variety of modes that the person may exhibit, depending on for example, a stress moment. The human behaviour may vary on an individual basis, including the effect of fatigue or decreased concentration on driving behaviour. But indeed the background reason may be of an underlying nature. The main cause can be related to the very highly complex technology needed to monitor the driver's behaviour and conditions, e.g., monitoring of the driver's eye activities that could have a tendency to affect the driver's behaviour.

There are, however, technical solutions available on the market that support the driver. A driver alert control system that detects the driver's behaviour, may consist of a camera for positioning the car on the road, in combination with a number of other sensors measuring the car behaviour and a control unit to estimate the driving path. The camera, normally continuously measures the distance between the car and the road lane markings. Additional information from the sensors detects the car's movements, e.g., acceleration, pedals and external parameters, e.g., side wind and road conditions. The computer then stores the information and calculates whether the driver is exihibiting an unstructured risk to lose control of the vehicle. If the risk is assessed as high, the driver is alerted via an audible signal and an additional text message appears on the car's information panel.

The illustrated alerting system that is now commercially built in vehicles is a first generation of complementary systems aiming to provide essential information to the driver. The supportive systems will most likely have a strong impact on the effectiveness of driving and most likely will also provide a major effect on reducing accidents and human suffering. But this problem-solving device may give rise to subsequent phenomenon that is not on the agenda today. An important question arises whether the driver may lower the general safety level and rely more on a system supporting the driver, than her own cognitive and sensing power. However, the alert system is, as just mentioned, a supportive system to the driver and is in no sense controlling the situation. The driver is still in charge of the situation and the sound/text information is considered to alert the driver to make a proper action, for example by braking the vehicle. The conclusion drawn in this type of symbiotic partnership is, the more sophisticated the system behaves, the more passive the driver becomes. We may experience drivers who trust that the system will take care of safety and will as a consequence drive more carelessly knowing that the system will make the expected alerts when neccessary.

The following question seems to arise as the technical consequences of designing more sophisticated and individual adapted artefacts in conjunction with the human perception.

In the long run, a continuous degradation of the human senses may occur, as we no longer have the need to actively make use of the sensing skills and strengthen our abilities to take care of our own capacity and by that it also means restricting the perceptual limitations, unless we are not activating our expectations onto new and adventures experiences.

This scenario may in some sense be frightening, and may in the future make us dependent on complementary artificial systems. However, this fact can on the other hand create opportunities to establish an understanding and awareness of the need to develop our sensing abilities. The understanding of the process of degradation, in the long term as during a life time seems to be of essential knowledge.

Indeed, the sensor abilities that can improve the understanding of a dynamic world will perform a vital function to increase human performance, namely in providing a development that most likely will change the human view of an increased added capacity. The individual solutions that increase, adapt or adjust complementary human abilities today exist in a range of areas. The era of a wooden leg, plastic hand or a glass eye have since some time now been substituted by bio-electronic solutions and high-tech materials. Nowadays, the field of research concentrates on completely new developments. Medical implants, tissue culture outside the body, bio-sensors and bio-materials that imitate the material of nature and is adapted by the body. Further, the direction of the develoment may be to examine into the field of bio-electronics and cells that are capable of restoring the internal human function of the damaged limb, organ or other parts of the body.

A future concept of science has emerged, to establish a market for designing artificial sensors that mimic the human perceptual sensing. For example, the electronic tongue, nose, ear and eye functions have been successful systems in complementing human ability. Also the ongoing attempts to design artificial skin that is able to sense from a certain proximity and perceive tactile information, will most likely create future improved qualities in prothesis design. The new generation of prosthesis have already started to use the knowledge gained from robotic science to make a complete set of artificial spare parts that in the long run is foreseen to be an extension of possibilities that at the end will have an impact when improving the human capability.

An illustrative example of a symbiotic device, is the completely new direction and upcoming trend of using jumping stilts as shown in Fig. 8.1. This new mobile performance provides humans with extraordinary abilities when it acts as an extension of the legs. The power of the jumping stilts can result in increased human abilities. The individual increase of performance is remarkable, to run at speeds of up to 35km/h, jump up to 2 metres high and take giant 3 metre strides. This performance is an example of a new trendy human symbiotic ability that would surely make an impression as a new possible transportation tool. The device will provide a new solution of an old habit, to make an extensive use of our own legs when transporting ourself, for example to or from work or simply taking a tour in the park.

This scenario will raise the question in the long run whether the replacement of spare parts, including additional sensor devices, will affect human evolution. How are humans relating to the development of intelligent perceptual increased devices that will be based upon external performance. When do we define natural

Figure 8.1. Super abilities with jumping stilts can indeed improve the individual mobility. Reprinted with permission from www.getjumpingstilts.com.

behaviour and when do we experience the increased capability as a peculiarity related to an additional value, in for example sports, employment, or family life?

The painting in Fig. 8.2, could with a touch of changeable complexity express a sense of greatness when referring to the following phrase:

> *"the presence of all that is present"*

> Martin Heidegger analysis of Nietzsches book Thus Spoke Zarathustra (1883) in his university lectures course 1951/1952, What is called thinking ?

Also, this will naturally give rise to the following question — when does the additional power from artificial parts become a human inconvenience?

This question naturally ends by raising the following issue;

- *who are you and what are you made of ?*

On the other hand, by looking in the future aspect of life and in that respect find basic arguments for future speculations, may then be considered a relatively easy task to come up with the right type of questions. However, when tackling this type of complex challenges, we surely need to take the past into consideration. The process of human development is bound to the history of the human evolutionary progress. The past knowledge may help us to understand the process of progress

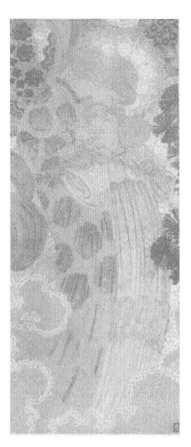

Figure 8.2. Human perception illustrated in a perceptual painting showing Augusto Giacometti, LaMusica (detail), 1899, Zurich, Switzerland, Kunsthaus. Reprinted with permission.

through evolution. The validation of the comments and arguments given in this book is with advantage in the hand of the reader, who may hopefully take further part in the discussions. Then the evolution will accordingly be pushed into further development in the direction of human benefit.

In the context of the past history that always will be connected to our development direction may be with advantage illustrated by the following painting. The painting, plaster and oil on paper, intends in this context to combine different perceptual modes. The beauty of the painting could be directed towards an illustration concerning the human perceptual complexity when combining music by visual art, as painted in the piece of art LaMusica shown in Fig 8.2.

Afterword

A dinner with some friends, a meal with the family or having fast food with the one you want to spend all your time with is an experience. Every meal has a message, communicates a feeling, to those who take part in it.

The substantial entirety of a meal is like art, seeking to inspire and to overshadow any feeling of "commodity" that it might have. Of course there are other possible perspectives on how, when and where we eat, but here I deal with the meal as an inspiration of impressions. Composing the experience is a matter of balancing perceptual sensing such as, points, lines, shapes, colours, proportions, movements, directions, light, sound and, orientation in the room and atmosphere and combining it with smell, taste and tactile scenarios, into an expressive and meaningful entirety of the impression we call the meal. In other words, the meaning of the meal emerges from the interplay of activating and balancing forces.

Which of our senses do we use in a fast food restaurant?

The unconsidered answer is "all of them" but the whole point of fast food: food as fuel naturally brings a stress and haste that covers and chokes our senses like cling film. One consequence of this may be a need to increase our ability to take in information through our senses. In principle, there is no limit to the efficiency of our senses, but a high throughput of information is a stress factor.

A meal is not only about the sense of taste. Consciously or unconsciously, diners experience all sorts of other sensory information: verbal and non-verbal communication with serving staff, the atmosphere of the room: light, colour, sound, directions and movements. There is also noise and body language from other guests, the odour of different bodies and the scent of food.

A pleasant, energy-giving and interesting meal depends upon a balance between all the factors involved. Our experience of a meal depends largely upon our sensory experiences: eyes, ears, nose, mouth, hand, mind and intuition all contribute. Creating the entirety of the meal involves the whole palette of "ingredients" that ultimately influence the diner's senses.

By combining the different disciplines in a meal, the Department of Culinary Arts at the University of Örebro, Sweden, bases its educational approach on five aspects of the meal: the food itself, room, the service encounter, the atmosphere and the economic control system. As a professor in art and design, as well as, being an artist and designer, I would like to extend this view.

From a designer's view of perspective, the meal can be defined in an odyssey designed to explore at least seven "ingredients" that make-up the meal's entirety, as I have developed in "The entirety of the meal: a designer's perspective"

(*Journal of Foodservice*, 24 Jan. 2008) into the following structure:

(1) the idea,
(2) art, balance and complexity
(3) line, shape and form
(4) light, colour and plates
(5) sound, music, rhythm, movement and atmosphere
(6) the food
(7) assembling the meal

The meal is clearly much more than mere food. The word meal is a symbol denoting a plethora of sensory and other meanings. We cannot smell the word meal or feel the flavour; it is just a symbol that summarises something much more complex and sensitive.

When these components are brought together in a balanced and dynamic structure, the good and interesting complexity may arise in an inexplicable, mysterious and experienced way. So much in life is straightened in our so-called civilised society, that can create stress and long-term illness. This is again a question of balance, between the material and the immaterial. In other words, an understanding of the possible perceptual impressions available and knowledge makes the freedom to create an entirety of the meal in new and interesting ways.

Enriching the individual's perceptual capabilities and making people to perceive more of an inviting meal is sincerely a way of wealth and added value to peoples' life. This creates new perspectives and unexpected possibilities.

This book will hopefully be an inspiration source for people communicating and influencing other people's perception. I would like to see a new evolution where we balance and make a more nuanced interaction between human senses, Nature and Technology.

As I see it, the development of Technology has to adapt and communicate Nature's way of structuring complexity. At that moment we have a golden opportunity to develop our lives, from all aspects, in a far more harmonic and healthy way.

Birgitta Watz
Prof. Art & Design, Sweden

References

[1] D. T. Ahmed, S. Shirmohammadi and J. C. Oliveira, Supporting large-scale networked virtual environments, *Proceedings of the IEEE Conference on Virtual Environments, Human-Computer Interfaces, and Measurement Systems*, Ostuni, Italy, June (2007).

[2] M. A. Alcántara, M. A. Artacho, J. C. González and A. C. Garcia, Application of product semantics to footwear design. Part I, Identification of footwear semantic-space applying differential semantics, *International Journal of Industrial Ergonomics*, **35**(8), August , 713–35, (2005a).

[3] M. A. Alcántara, M. A. Artacho, J. C. González and A. C. Garcia, Application of product semantics to footwear design. Part II, Comparison of two clog designs using individual and compared semantic profiles, *International Journal of Industrial Ergonomics*, **35**(8), 713–735, (2005).

[4] W. T. Anderson, Augmentation, symbiosis, transcendence; technology and the future(s) of human identity, *Futures*, **35**(5), June, 535–546, (2003).

[5] A. Anund, B. Peters and G. Kecklund, Fatigued driving-drivers' point of view, VTI-Report, No. 498, English summary (2004).

[6] M. Augoustinos, I. Walker and N. Donaghue, *Social Cognition, An Integrated Introduction*, SAGE (2006).

[7] R. Bajcsy, Active perception vs. Passive perception, *Proceedings of the 3rd Workshop on Computer Vision*, Washington DC, IEEE Press, 55–59, 1985.

[8] R. Bajcsy, Active perception, *Proceedings of IEEE*, **7**, 996–1005, (1988).

[9] S. Ballestero, S. Reales, J. M. De Leon and B. Garcia, The perception of ecological textures by touch: does the perceptual space change under bimodal visual and haptic exploration?, *Eurohaptics Conference 2005*, March, 635–638, (2005).

[10] L. M. Bartoshuk, Genetic and pathological taste variation: What can we learn from animal models and human disease?, *Ciba Found Symposium*, **179**, 251–262, (1993).

[11] L. M. Bartoshuk, V. B. Duffy, D. Reed and A. Williams, Supertasting, Earaches and head injury: Genetics and pathology alter our taste worlds, *Neuroscience and Biobehavioral Reviews*, **20**(1), 379–387, (1996).

[12] V. Batagelj, H. H. Bock, A. Ferligoj and A. Ziberna, (eds.), Data science and classification, In: *Studies in Classification, Data Analysis, and Knowledge Organization*, Springer (2006).

[13] D. T. Batarseh, T. N. Burcham, G. M. McFadyen, An ultrasonic ranging system for the blind (1997).

[14] F. Bear, B. Connors and M. Paradiso, *Neuroscience Exploring the Brain*, William and Wilkins (1996).

[15] M. R. Benjamin, J. A. Curcio and P. M. Newman, Navigation of unmanned marine vehicles in accordance with the rules of the road, *Proceedings of the 2006 Conference on International Robotics and Automation*, 15–19 May, Orlando, USA, (2006).

[16] K. Bergstrand, K. Carlsson, P. Wide and B. Lindgren, Sound detection in noisy environment-locating drilling sound by using an artificial ear, *ROSE 2004 International Workshop on Robot Sensing*, 55–60, (2004).

[17] G. Berlucchi, V. Moro, C. Guerrini and S. M. Aglioti, Dissociation between taste and tactile extinction on the tongue after right brain damage, *Neuropsychologia*, **42**(8), 1007–1016, (2004).

[18] J. B. Best, Cognitive psychology, in De Greene (ed.), *Systems Psychology*. Mc Graw-Hill, New York, (1993).

[19] L. Biel, *Modeling of Perceptual Systems — A Sensor Fusion Model with Active Perception*, PhD thesis, Örebro University (2002).

[20] L. Biel and P. Wide, Active perception in a sensor fusion model, *Sensor Fusion: Architectures, Algorithms and Applications, VI, Proc.*, **4731**, 164–175, (2002a).

[21] L. Biel and P. Wide, Active perception for autonomous sensor systems: an emerging paradigm, *Proceedings of the IEEE Instrumentation & Measurement Magazine*, 27–30, (2000).

[22] A. Billard, S. Calinon, R. Dillmann and S. Schaal, Robot programming by demonstration, in Siciliano, B., Khatib, O. (eds.), *Handbook of Robotics*, Springer, (2008).

[23] Biosensor Applications, www.biosensor.se/technology, February (2009).

[24] D. C. Blanchaard, R. J. Blanchard and J. Rosen, *Editorial Olfaction and Defence, Neuroscience and Biobehavioral Reviews*, **32**, 1207–1208, (2008).

[25] E. P. Blasch and S. Plano, JDL Level 5 fusion model: user refinement issues and applications in group tracking, *SPIE*, 4729, Aerosense, 270–279, (2002).

[26] J. Blauert, Partly translated by J. Allen, *Spatial Hearing, the Psychophysics of Human Sound Localisation*, MIT Press (1996).

[27] A. H. Bloksma and W. Bushuk, Rheology and the chemistry of dough, wheat chemistry and technology, in Pomeranz, Y. (ed.), *American Association of Cereal Chemists*, St. Paul, Minnesota, USA, **58**, 481, (1988).

[28] E. Boivin and I. Sharf, Optimum grasp planner and vision-guided grasping using a three-finger hand, *Journal of Industrial Robot*, **32**(1), 35–42, (2005).

[29] H. H. Bothe, M. Persson, L. Biel and M. Rosenholm, Multivariate sensor fusion by a neural network model, *Proceedings of Fusion'99*, Sunnyvale USA, 1094–1101, (1999).

[30] W. Bourgeois, G. Gardey, M. Servieres and R. M. Stuetz, A chemical sensor array based system for protecting wastewater treatment plants, sensors and actuators, B **91**(1–3), 109–116, (2003).

[31] M. Broxvall, M. Gritti, A. Saffiotti, B. S. Seo and Y. J. Cho, PEIS ecology: Integrating robots into smart environments, *Proceedings of the IEEE International Conference on Robotics and Automation, ICRA06*, 212–218, (2006).

[32] S. Calinon, *Continuous Extraxtion (Author: Extraction?) of Task Constraints in a Robot Programming by Demonstration Framework*, PhD thesis, Ecole Polytechnique Federale de Lausanne, Switzerland (2007).

[33] H- W. Chen, R- J. Wu, H- H. Chen, C -Y. Liu and C- H. Yeh, The application of conductivity on the electronic tongue, *Proceedings of the IEEE International Workshop on Cellular Neural Networks and their Applications, CNNA*, 19–22, (2005).

[34] G. D. Christian, *Analytical Chemistry*, 6th ed., John Wiley and Sons, Inc (2004).

[35] P. R. Cohen, *Empirical Methods for Artificial Intelligence*, MIT Press, (1995).

[36] C. Conrad, Esoteric or exoteric? Music in medicine, *Medscape Journal of Medicine*, **0**(1), 28 January, 20, (2008).

[37] M. Cooley, On human-machine symbiosis, in cognition, communication and interaction, in Gill, S. P. (ed.), *Human-Computer Interaction Series*, Springer London, 457–485, (2008).

[38] K. Dalamagkidis, K. P. Valavanis and L. A. Piegl, On unmanned aircraft systems issues, challenges and operational restrictions preventing integration into the National Airspace System, *Progress in Aerospace Sciences* **44**(7-8), October/November, 503–519, (2008).

[39] A. Damasio, The feeling of what happens. *Body and Emotion of Consciousness*, A Harvest Book Harcourt, Inc., London, UK, (1999).

[40] A. Damasio, *Descartes' Error. Emotion, Reason and the Human Brain*, Penguin Books, London, UK, (1994).

[41] E. Daniel, Noise and hearing loss: A review, *Journal of School*, Health, **77**(5), 225–231, (2007).

[42] A. K. Das, G. Kantor, R. V. Kumar and G. A. S. Pereira, R. Peterson, D. Rus, S. Singh and J. Spletzer, Distributed search and rescue with robot and sensor teams, *Proceedings of the 4th International Conference on Field and Service Robotics*, USA, (2003).

[43] N. Debnath, Z. A. Hailani, S. Jamalundin and S. A. K. Aljudin, An electronically guided walking stick for the blind, *Proceedings of the 23rd Annual EMBS International Conference*, October, Istanbul, Turkey, (2001).

[44] A. J. DeCasper and W. P. Fifer, Of human bonding: newborn prefer their mothers' voice, *Science*, **208**(4448), June, 1174–1176, (1980).

[45] A. K. Deisingh, D. C. Stone and M. Thompson, Applications of electronic noses and tongues in food analysis, *International Journal of Food Science and Technology*, **39**, 587–604, (2004).

[46] E. Demeester, A. Huntemann, D. Vanhooydonk, G. Vanacker, H. Van Brussel and M. Nuttin, User-adapted plan recognition and user-adapted shared control: A Bayesian approach to semi-autonomous wheelchair driving, *Autonomous Robots*, **24**(2), 193–211, (2008).

[47] P. M. A. Desmet, Measuring emotion; development and application of an instrument to measure emotional responses to products, in Blythe, M. A. Monk, A. F. Overbeeke, K. Wright, P. C., (eds.), *Funology: From Usability to Enjoyment*, Kluwer Academic Publishers, Dordrecht, 111–123, (2003).

[48] D. Dubois, Categories as acts of meaning: the case of categories in olfaction and audition, *Cognitive Science Quaterly*, **1**, 35–68, (2000).

[49] D. A. Edwards K. T. Griffis and C. Tardivel, Olfactory bulb removal effects on sexual behavior and partner preference in male rats, *Physiology & Behavior*, **48**(3), 447–450, (1990).

[50] J. S. A. Edwards and I. B. Gustafsson, The five aspects meal model, *Journal of Food Service*, **19**, 4–12, (2008).

[51] M. R. Endsley, Theoretical underpinnings of situation awareness: A critical review, in Endsley, M. R. Garland, D. L. (eds.) *Situation Awareness Analysis and Measurement*, Lawrence Erlbaum Associates (2000).

[52] K. Erlandsson, A. Dsilna, I. Fagerberg and K. Christensson, Skin-to-skin care with the father after cesarean birth and its effect on newborn crying and prefeeding behavior, *Birth*, **34**(2), June, 105–114, (2007).

[53] K. H. Esbensen, *Multivariate Data Analysis — In Practice*, 3th ed, CAMO, Oslo Norway (2000).

[54] M. W. Eysenck, *Cognitive Psychology*, Lawrence Erlbaum Associates, London, UK, (1993).

[55] M. W. Eysenck and M. T. Keane, Chapter 4: Theories of perception, movement, and action, in *Cognitive Psychology: A Student's Handbook*, Lawrence Erlbaum, Hillsdale, USA, (1995).

[56] R. Feldhoff, C. Saby and P. Bernadet, Detection of perfumes in diesel fuels with semiconductor and mass-spectrometry based electronic noses, *Flavour and Fragrance Journal*, **15**(4), 215–222, (2000).

[57] Figaro, http://www.figaro.co.jp (2009).

[58] M. S. Fisher, *Software Verification and Validation*, Springer, (2007).

[59] G. Forsgren, *Evaluation of Gas Sensors for Monitoring Volatile Emitted Packaging Board Products*, Lic. Thesis,Linkoping University, Sweden, (1999).

[60] J. Fraden, *AIP Handbook of Modern Sensors, Physics, Designs and Applications*, American Institute of Physics, (1993).

[61] W. J. Freeman, *Comparison of Brain Models for Active vs. Passive Perception, Information-Sciences*, 97–107, (1999).

[62] U. Frykman, E. Hedborg, A. Spetz, H. Sundgren, S. Welin and F. Winquist, Artificial 'olfactory' images from a chemical sensor using a light-pulse technique, *Nature*, **352**(6330), 47, (1991).

[63] J. C. Fu, C. A. Troy and P. J. Phillips, A matching pursuit approach to small drill bit breakage prediction, *International Journal of Production Research*, **37**(14), 3247–3261, (1999).

[64] M. C. Gacula, *Descriptive Sensory Analysis in Practice*, Blackwell Publishing (1997).

[65] M. A. Garces, Infrasonic signals generated by volcanic eruptions, *Proceedings of Geoscience and Remote Sensing Symposium, IGARSS*. (2000).

[66] J. Gardner and Barlett, *Electronic Noses, Principles and Applications*, Oxford University Press, New York, USA (1999).

[67] G. A. Gescheider, Some comparisons between touch and hearing, *IEEE Transaction on Man Machine Systems*, **11**(1), 28–35, (1970).

[68] J. J. Gibson, *The Ecological Approach to Visual Perception*, Lawrence Erlbaum Associates, (1987).

[69] K. S. Gill, *Human Machine Symbiosis*, Springer, (1996).

[70] A. L. Glass and K. J. Holyoak, *Cognition*, 2nd ed., McGraw-Hill, Inc (1986).

[71] W. Goepel and G. Reinhardt, Metal oxide sensors: New devices through tailoring interfaces on the atomic scale, *Sensors Update*, **1**(1), 49–120, (2001).

[72] E. B. Goldstein, *Sensation and Perception*, Cengage Learning (2006).

[73] D. M. Gordon, *Ants at Work: How an Insect Society is Organized*, W.W. Norton & Company, Inc., New York, (1999).

[74] X. S. Guo, Y. Q. Chen, X. L. Yang and L. R. Wang, Development of a novelelectronic tongue system using sensor array based onpolymer films for liquid phase testing, *Proceedings of the 27th Annual International Conference on the IEEE Engineering in Medicine and Biology Society*, Shanghai, China, 4, (2005).

[75] D. L. Hall and J. Llinas, (ed.) *Handbook of Multisensor Data Fusion*, CRC Press (2001).

[76] D. L. Hall, *Mathematical Techniques in Multisensor Data Fusion*, Artech House, Boston, USA, (1992).

[77] P. Harmo, T. Taipalus, J. Knuuttila, J. Vallet and A. Halme, Needs and solutions — home automation and service robots for the elderly and disabled, in *Proceedings of the IEEE/RSJ International Conference on Intelligent Robots and Systems*, Edmonton Canada, 1 (2005).

[78] M. He, J. Zeng and Y. Liu, *et al.*, (Author: to provide all names) Refractive error and visual impairment in urban children in southern China, *Investigative Ophthalmolology & Visual Science*, 45 793–799 (2004).

[79] M. Heidegger, (1951/1952). *What is Called Thinking*, translated by J. Glenn Gray, Harper & Row Publishers, New York, (1968).

[80] H. Hirano and F. Makota, *JIT is Flow: Practice and Principles of Lean Manufacturing*, PCS Press, Inc (2006).

[81] K. Holmberg, (ed.), *Handbook of Applied Surface and Colloid Chemistry*, FÖRLÄGGARE? 1, (2002).

[82] R. C. Hoseney, *Bread Baking, Cereal Foods World*, 39, 180, (1994).

[83] J. H. Huijsing and in G. Meijer, (ed.), *Smart Sensor Systems*, Wiley Blackwell, (2008).

[84] A. Huntemann, E. Demeester, M. Nuttin and H. Van Brussel, Online user modeling with Gaussian processes for Bayerian plan recognition during power-wheelchair steering, in *IEE/RSJ International Conference on Intelligent Robots and Systems, IROS*, 285–292, (2008).

[85] B. Iliev, M. Lindquist, L. Robertsson and P. Wide, A fuzzy technique for food- and water quality assessment with an electronic tongue, *Fuzzy Sets and Systems*, 157, 1155–1168, (2006).

[86] J. Ip, S- M. Saw, K. Rose, I. G. Morgan, A. Kifley, J. J. Wang and P. Mitchell, Role of near work in myopia: findings in a sample of Australian school children, *Investigative Ophthalmolology & Visual Science*, 49(7), 2903–2910, (2008).

[87] N. Ausovec, K. Jausovec and I. Gerlic, The influence of Mozart's music on brain activity in the process of learning, *Clinical Neurophysiology*, 17(12), December, 2703–2714, (2006).

[88] J. S. Jenkins, The Mozart effect, *Journal of the Royal Society of Medicine*, 94(4), April, 170–172, (2001).

[89] T. Jindo and K. Hirasago, Application studies to car interior of Kansei engineering, *International Journal of Industrial Ergonomics*, 19, 105–114, (1997).

[90] J. Jouper and P. Hassmén, Intrinsically motivated Qigong exercisers are more concentrated and less stressful, *American Journal of Chinese Medicine*, **36**(6), 1051–1060, (2008).

[91] J. Jouper, P. Hassmén and M. Joansson, Qigong exercise with concentration predicts increase health, *American Journal of Chinese Medicine*, **34**(6), 949–957, 2006.

[92] B. Kalyanaraman, D. M. Supp and S. T. Boyce, Medium flow rate regulates viability and barrier function of engineered skin substitutes in perfusion culture, *Tissue Engineering, Part A*, **14**(5), 583–593, (2008).

[93] M. A. Kenna, Music to your ears: is it a good thing?, *Acta Paediatrica*, **97**, 151–152, (2008).

[94] A. Kikukawa, S. Yagura and T. Akamatsu, A 25-year prospective study of visual acuity in the Japan Air Self Defence Force personnel, *Aviation Space and Environmental Medicine*, **70**(5), 447–450, (1999).

[95] A. K. Klein, *Sensor and Data Fusion Concepts and Applications*, 2nd ed., SPIE Optical Engineering Press (1999).

[96] M. Kotani, T. Arimoto, S. Ozawa and K. Akazawa, *Application of Independent Component Analysis to Detection of Gas Leakage Sound*, Kobe University, Japan, 2287–2291, (2001).

[97] D. Kroenke and D. Auer, *Database Processing: Fundamentals, Design, and Implementation*, 10th ed, Prentice Hall, (2005).

[98] D. Kroenke and D. Auer, *Database Processing*, 11th ed., Prentice Hall, (2009).

[99] A. Legin, D. Kirsanov, A. Rudnitskaya, J. J. L. Iversen, B. Seleznev, K. H. Esbensen, J. Mortensen, L. P. Houmoller and Y. Vlasov, Multicomponent analysis of fermentation growth media using the electronic tongue (ET), *Talanta*, **64**, 766–772, (2004).

[100] H. T. Lawless and H. Heymann, *Sensory Evaluation of Food, Principles and Practices*, Chapman & Hall, (1999).

[101] K. Li, S. Thompson, P. A. Wieringa and J. X Peng, A study on the role of human operators in supervised automation system and its implications, *Proceedings of the 4th World Congress on Intelligent Control and Automation*, **4**, 3288–3293, (2002).

[102] Y. Li, Perception of temperature, moisture and comfort in clothing during environmental transient, *Ergonomics*, **48**(3), February, 234–248, (2005).

[103] Licklider, Man-computer symbiosis, *IRE Trans. on Human Factors in Electronics*, HFE-1, March, 4–11, (1960).

[104] A. J. Lilienthal, A. Loutfi and T. Duckett, Airborne chemical sensing with mobile robots. *Sensors*, **6**, 1 October, 1616–1678, (2006).

[105] M. Lindquist and P. Wide, New sensor system for drinking water quality, in *Proceedings of the Sensors for Industry Conference*, New Orleans, USA, 30–34, (2004).

[106] M. Lindquist, *Electronic Tongue for Water Quality Assessment*, PhD thesis, Örebro University, Sweden, (2007).

[107] J. Llinas and D. Hall, An introduction to multi-sensor data fusion, in *Proceedings of the IEEE*, **85**(1), January, (1997).

[108] E. P. Lopes, Application of a blind person strategy for obstacle avoidance with the use of potential fields,in *Proceeds off the IEEEICRA*, Seoul. Korea, May, 2911–2916, (2001).

[109] A. Loutfi and P. Wide, Symbolic estimation of food odours using fuzzy techniques, in *Proceedings of the Information Processing and Management of Uncertainty, IPMU 2002*, Annecy, France, 919–926, (2002).

[110] A. Loutfi, J. Widmark, E. Wikstrom and P. Wide, Social agent: Expressions driven by an electronic nose, in *VECIMS 2003, International Symposium on Virtual Environments, Human-Computer Interfaces, and Measurement Systems*, Lugano Switzerland, 95–100, (2003).

[111] A. Loutfi, S. Coradeschi, L. Karlsson and M. Broxvall, Putting olfaction into action: using an electronic nose on a multi-sensing mobile robot, in *Proceedings of the IEEE/RSJ International Conference on Intelligent Robots and Systems (IROS 2004)*, Sendai, Japan, 347–352, (2004).

[112] A. Loutfi, M. Lindquist, P. Wide, M. Ilyas and I. Mahgoub, (eds), *Handbook of Sensor Networks: Compact Wireless and Wired Sensing Systems,*, CRC Press (2005).

[113] A. Loutfi, *Odour Recognition using Electronic Noses in Robotic and Intelligent Systems*, PhD thesis, Örebro University, Sweden, (2006).

[114] I. Lundström, Hydrogen sensitive MOS-structures, Part I: principles and applications, *Sensors and Actuators*, **1**, 403–426, (1981).

[115] M. Manske and G. Cordua, Understanding the Sommelier effect, *International Journal of Contemporary Hospitality Management*, **17**(6–7), 569–576, (2005).

[116] Y. Matsubara and M. Nagamachi, Hybrid Kansei engineering system and design support, *International Journal of Industrial Ergonomics*, **19**, 81–92, (1997).

[117] G. Meijer and in G. Meijer, (ed.), *Smart Sensor Systems*, Wiley Blackwell, (2008).

[118] Meilgaard, M., Civille, G. V., Carr, T. *Sensory Evaluation Techniques*, CRC Press (2007).

[119] J. C. Merrit, S. Game, O. D. William, D. Blake, Visual acuity in preschool children: the Chapel Hill-Durham day-care vision study, *Journal of the National Medical Association*, **88**, 709–712, (1996).

[120] C. Mikkelsen, *Information Visualization of N-dimensional Vector Spaces*, MSc thesis 2004, Linköping University, Sweden (2004).

[121] M. Minnaert, *The Nature of Light and Colour in the Open Air*, Dover Publications Inc., New York, **113**, (1954).

[122] J. Mojet, J. Heidema, E. Christ-Hazelhof, Effect of concentration on taste-taste interactions in food for elderly and young subjects, *Chemical Senses*, **29**(8), 671–681, (2004).

[123] T. C. Morata, Young people: their noise and music exposures and the risk of hearing loss, *International Journal of Audiology*, **46**, 111–112, (2007).

[124] Y. Mukaibo, H. Shirado, M. Konyo and T. Maeno, Development of a texture sensor emulating the tissue structure and perceptual mechanism of human fingers, in *Proceedings of the 2005 IEEE International Conference on Robotics and Automation, ICRA 2005*, April, Barcelona, Spain, 2565–2570, (2005).

[125] T. Naes and E. Risvik, *Multivariate Analysis of Data in Sensory Science, Elsevier Science B.V.*, (1996).

[126] M. Nagamachi, *The Story of Kansei Engineering*, Kaibundo Publishing, Tokyo, Japan, (1995).

[127] K. S. Naidoo, A. Raghunandan and K. P. Mashige, *et al.* (Author: please provide all names) Refractive error and visual impairment in African children in South Africa, Invest. *Ophthalmology and Visual Science*, **44**, 3764–3770, (2003).

[128] H. Nanto and R. Stetter, *Handbook of Machine Olfaction: Electronic Nose Technology, (Chapter – Introduction to chemosensors)*, Wiley-VCH, 79–104, (2003).

[129] J. Neitz, J. Carroll and M. Neitz, Color vision, almost reason enough for having eyes, *Optics & Phtonics News*, January (2001).

[130] A. Nelson, W. Hartl, K. W. Jauch, G. L. Friccione, H. Benson, A. L. Warshaw and C. Conrad, The impact of music on hypermetabolism in critical illness, *Curr. Opin. Clin. Metab. Care*, **11**(6), November, 790–794.

[131] B. Y. Newman, Drowsy driving, the silent killer, *Journal of the American Optometric Association*, **75**(8), 481–482, (2004).

[132] F. Nietzsche, *Thus Spoke Zarathustra: A Book for All and None*, translated by Kaufmann, W., The Modern Library, New York, (translated in 1995).

[133] H. Nomura, F. Ando, N. Niino, H. Shimokata and Y. Miyake, Age-related change in contrast sensivity among Japanese adults, *Japanese Journal of Ophthalamology*, **47**(3), 299–303 (Cochrane collaboration, 2007), (2003).

[134] T. Norretranders, *The User Illusion: Cutting Consciousness Down to Size*, Penguin Schiffman, H. R. Press Science, (1998).

[135] J. Olsson, P. Ivarsson and F. Winquist, Determination of detergents in washing machine wastewater with a voltammetric electronic tongue, *Talanta*, **76**(1), June, 91–95, (2008).

[136] C. E. Osgood, G. J. Suci and P. H. Tannenbaum, *The Measurement of Meaning*, University Illinois Press, Illinois, USA, (1957).

[137] S. E. Palmer, Modern theories of Gestalt perception, *Journal of Mind & Language*, **5**(4), 289–323, (1990).

[138] G. Pavlin, M. G. Maris and F. C. A. Groen, Causal Bayesian networks for robust and efficient fusion of information obtained from sensors and humans, in *Proceedings of the Instrumentation and Measurement Technology Conference*, Warsaw, Poland, 1–6, (2007).

[139] T. C. Pearce, J. W. Gardner, S. S. Schiffman and H. T. Nagle, *Handbook of Machine Olfaction*, Wiley-VCH, (2003).

[140] A. Perera, A. Gomez-Baena, T. Sundic, T. Pardo and S. Marco, Machine olfaction: Pattern recognition for the identification of aromas, in *Proceedings of the International Conference on Pattern Recognition*, **16**(2), 410–413, (2002).

[141] E. M. Petriu, T. E. Whalen, I. J. Rudas, D. C. Petriu and M. D. Cordea, Human-instrument partnership for multimodal environment perception, in *Proceedings of the IEEE International Instrumentation and Measurement Technology Conference*, May, Victoria Vancouver Island, Canada, 1263–1268, (2008).

[142] T. P. L. Quek, C. G. Chua, C. S. Chong, J. H. Chong, H. W. Hey, J. Lee, Y. F. Lim and S -M. Saw, *J. Ophthalmatic. Opt*, (Author: please provide complete journal name), **24**, (1), 47–55, (2004).

[143] F. H. Rauscher, G. L. Shaw and K. N. Ky, Music and spatial task performance, *Nature*, **365**, 611, (1993).

[144] R. A. Rensink, Visual sensing without seeing, *Psychological Science*, **15**(1), January, 27–32, (2004).

[145] C. Reynolds, A. Cassinelli, Y. Watanabe and M. Ishikawa, *Manipulating Perception, 6th European Conference on Computing and Philosophy*, June, Montpellier, France, (2008).

[146] G. Ristovska, D. Gjorgjev and N. Pop Jordanova, Psychosocial effects of community noise: cross sectional study of school children in urban center of Skopje, Macedonia, *Croatian Medical Journal*, **45**(4), 473–436, (2004).

[147] D. J. Risovic, K. R. Misailovic, J. M. Eric-Marinkovic, N. G. Kosanovic-Jakovic, S. M. Milenkovic and L. Z. Petrovic, *European Journal of Ophthalmology*, **18**(1), 1–6, (2008).

[148] D. Robaei, K. Rose, E. Ojaimi, A. Kifley, S. Huynh and P. Mitchell, Visual acuity and the cause of visual loss in a population-based sample of 6-year-old Australian children, *American Academy of Ophthalmology*, **112**(7), 1275–1282, (2005).

[149] L. Robertsson and P. Wide, Improving food quality analysis using a wavelet method for feature extraction, in *Proceedings of IMTC 2005 Instrumentation & Measurement Technology Conference*, May, Ottawa, Canada, (2005).

[150] L. Robertsson, B. Iliev, R. Palm and P. Wide, Perception modeling for human-like artificial sensor systems, *International Journal of Human-Computer Studies*, **65**, 446–459, (2007).

[151] L. Robertsson, *Perception Modeling and Feature Extraction for an Electronic Tongue*, PhD thesis, Örebro University, Sweden, (2007).

[152] G. Robles De La Torre, Principles of haptic perception in virtual environments, in Grunwald, M. (ed.), *Human Haptic Perception*, Birkhäuser Verlag, 363–379, (2008).

[153] K. R. Rogers and A. Mulchandani, *Affinity Biosensors: Techniques and Protocols*, Humana Press, (1998).

[154] G. J. Romanes, *Animal Intelligence*, BiblioBazaar, LLC, (2008).

[155] J. Ronnberg, E. Samuelsson and E. Borg, Exploring the perceived world of the deaf-blind: on the development of an instrument. *International Journal of Audiology*, March, **41**(2), 136–143, (2002).

[156] K. A. Rose, I. G. Morgan, J. Ip, A. Kifley, S. Huynh, W. Smith and P. Mitchell, Outdoor activity reduces the prevalence of myopia in children, *Journal of Ophthalmology*, **115**(8), 1279–1285, (2008).

[157] N. E. Rosenthal, *Winter Blues: Seasonal Affective Disorder: What Itt Is and How to Overcome It*, The Guilford Publication Inc., New York, (1998).

[158] A. Saffiotti, M. Broxvall, M. Gritti, K. LeBlanc, R. Lundh, J. Rashid, B. S. Seo and Y. J. Cho, The PEIS-Ecology Project: vision and results, in *Proceedings of the IEEE/RSJ International Conference on Intelligent Robots and Systems (IROS)*, Nice, France, 2329–335, (2008).

[159] J. Sandsten, H. Edner and S. Svanberg, Gas visualization of industrial hydrocarbon emissions, *Optical Express*, **12**, 1443–1451, (2004).

[160] J. Sandsten, *Development of Infrared Spectroscopy Techniques for Environmental Monitoring*, PhD thesis, Lund University, Sweden, (2004a).

[161] SBU, The Swedish Council on Technology Assessment in Health Care Report Light therapy for depression, and other treatment of seasonal affected disorder, revision of Chapter 9 in SBU Report Treatment of Depression 2004, No. 166/2 (2007).

[162] S. S. Schiffman, Perception of taste and smell in elderly persons, *Crit. Rev. Food Sci. Nutr.*. **33**(1), 17–26, (1993).

[163] H. R.Schiffman, *Sensation and Perception — An Integrated Approach*, 4th. ed., John Wiley & Sons, Inc (1996).

[164] B. N. Schilit, N. L. Adams and R. Want., Context aware computing applications, in *Proceedings of the Workshop on Mobile Computing Systems and Applications*, Santa Cruz, USA, December, IEEE Computer Society, (1994).

[165] A. Scozzari and P. Wide, The process from a redundant and general sensor concept-towards an optimal sensor strategy for the assessment of drinking water quality, in *Proceedings of the International Instrumentation and Measurement Technology Conference*,Victoria, Canada, May, 836–841, (2008).

[166] F. C. P. Sebelius, B. N. Rosen and G. N. Lundborg, Refined myoelectric control in below-elbow amputees using artificial neural network and a data glove, *Journal of Hand Surgery (A)*, **30**(4), 780–789, (2005).

[167] H. Shirado and T. Maeno, Modeling of human texture perception for tactile displays and sensors, in *Eurohaptics Conference 2005*, March, 629–630 (2005).

[168] M. Siegel, The sense-think-act paradigm revisited, in the *International IEEE Workshop on Robotic Sensing ROSE 2003*, 1–5, 1954, (2003).

[169] D. A. Skoog, D. M. West, F. J. Holler and S. R. Crouch, *Fundamentals of Analytical Chemistry*, Thomson-Brooks/Cole, 8th ed, (2004).

[170] A. Slater and V. Morison, Shape constancy and slant perception at birth, *Perception*, **14**(3), 337–344, (1985).

[171] A. Smailagic, D. P. Siewiorek, J. Anhalt, F. Gemperie, D. Salber, S. Weber, J. Beck and J. Jennings, Towards context aware computing: experiences and lessons, *IEEE Journal on Intelligent Systems* **26**(3), June, 38–46, (2001).

[172] J. Song, J. Zhou and Z. L. Wang, Piezoelectric and semiconducting coupled power generating process of a single ZnO belt/wire. A technology for harvesting electricity from the environment, *Nano Letters*, **8** August, 1656–1662, (2006).

[173] A. I. Spielman, Chemosensory function and dysfunction, *Critical Reviews in Oral Biology & Medicine*, **9**(3), 267–291, (1998).

[174] N. A Stillins *et al.*, *Cognitive Science*, MIT Press, London (1995).

[175] T. Sundic, S. Marco, J. Samitier and P. Wide, Electronic tongue and electronic nose data fusion in classification with neural networks and fuzzy logic based models, in *Proceedings of the 17th IEEE Instrumentation and Measurement Technology Conference 2000*, 1474–1479, (2000).

[176] J. K. Sung H. Young and M. Kim, Hyuk, *Development of an Intelligent Guide-Stick for the Blind, Proceedings of the 2001 IEEE International Conference on Robotics & Automation*, May, Korea, (2001).

[177] J. Tegin and J. Wikander, Tactile sensing in intelligent robotic manipulation: A review, *Industrial Robot*, **32**(1), 64–70, (2005).

[178] S. Theodoridis and K. Koutroumbas, *Pattern Recognition*, 3rd ed., Academic Press, San Diego, USA (2006).

[179] M. Thuillard, *Wavelets in Soft Computing*, World Scientific Publishing Co. Pte. Ltd (2001) Singapore.

[180] C. Törngren and G. Fodor, 3D viewing of model sensitivity properties in MIMO controllers, in *2003 IEEE International Symposium on Virtual Environments, Human-Computer Interfaces and Measurement Systems*, p (Author: page number missing) Toyota (2009), http://www.toyota.co.jp/en/special/robot/ (2003).

[181] C. Turati, V. Macchi Cassia, F. Simion and I. Leo, Newborns' face recognition: role of inner and outer facial features, *Child Development*, March-April, **77**(2), 297–311, (2006).

[182] C. Urmson and W. Whittaker, Self-driving cars and the urban challenge, *IEEE Intelligent Systems*, **23**(2), March–April, 66–68, (2008).

[183] J. Van den Stock, R. Righart and B. de Gelder, **7**(3), 487–494, 2007.

[184] J. J. Valdés, *Virtual Reality Representation of Information Systems and Decision Rules: An Exploratory Tool for Understanding Data and Knowledge, Artificial Intelligence LNAI 2639*, Springer-Verlag, 615–618, (2003).

[185] M. Valera and S. A. Velastin, *Proceedings of Vision, Image and Signal Processing*, **152**(2), April, 192–204, (2005).

[186] B. M. Velichkovsky, A. Rothert, M. Kopf, S. M. Dornhöfer and M. Joos, Towards an express diagnostics for level of processing and hazard perception, *Transportation Research Part F: Traffic Psychology and Behaviour. Special Issue: Eye movements, Visual Attention and Driving Behaviour*, **5**(2), Jun, 145–156, (2002).

[187] S. Vitale, M. F. Cotch, R. Sperduto and L. Ellwein, Costs of refractive correction of distance vision impairments un the United States, 1999-2002, *Ophthalmology*, **112**(12), 2163–2170, (2006).

[188] J. Wang, *Analytical Electrochemistry*, Wiley-VCH, (1994).

[189] B. Wats, The entity of the meal: A designer's perspective, *Journal of Food Service*, **19**, 86–104, (2008).

[190] B. Wenzel, *Handbook of Perception, Vol. III: Biology of Perceptual Systems*, (eds.) E. Carterette and M. Friedman, Academic Press, New York, (1973).

[191] B. Winn, D. Whitaker, D. B. Elliot and N. J. Phillips, Factors affecting light-adapted pupil size in normal human subjects, *Investigative Ophthalmology & Visual Science*, **35**(3), March, 1132–1137.

[192] P. Wide, New knowledge based measurement methods in the bread baking process, in *Proceedings of the 9th World Congress of Food Science and Technology*, Budapest, Hungary, **116**, (1995).

[193] P. A. Wide, human-knowledge-based sensor implemented in an intelligent fermentation-sensor system, *Sensors and Actuators B*, **32**, 227–231, (1996).

[194] P. Wide, A human perception related chaos based sensor fusion system for on-line supervisory control, in *IEEE Instrumentation & Measurement Technology Conference*, Brussels, Belgium, 1402–1406, 1996a.

[195] P. Wide, F. Winquist and D. Driankov, An air-quality sensor system with fuzzy classification, *Measurement Science Technology*, **8**, 138–146, (1997).

[196] P. Wide, The human perception based sensor system, in *IEEE Workshop on Emerging Technologies, Intelligent Measurement and Virtual Systems for Instrumentation and Measurement, ETIMVIS'98*, St. Paul, USA, 94–103, (1998a).

[197] P. Wide, F. Winquist, P. Bergsten and E. M. Petriu, Human-based multi-sensor fusion method for artificial nose and tongue sensor data, in *IEEE Transaction of Instrumentation & Measurement Technology*, **47**(5), October, 1998.

[198] P. Wide, I. Kalaykov and F. Winquist, The artificial sensor head: A new approach in assessment of human based quality, in *Proceedings of the Second International Conference on Information Fusion*, July, 1144–1149, (1999).

[199] P. Wide, The human decision making in the dough mixing process estimated in an artificial sensor system, *Journal of Food Engineering*, **39**, 39–46, (1999a).

[200] P. Wide and S. Asp, The electronic head: A virtual quality instrument, *Proceedings of the IEEE International Conference on Industrial Electronics, Control and Instrumentation*, October **98**, (2000).

[201] P. Wide, The electronic head: a virtual quality instrument, *IEEE Transactions on Industrial Electronics*, **48**(4), August, 766–769, (2001).

[202] P. Wide, nov New Zealand (2007).

[203] J. S. Wilson, *Sensor Technology Handbook*, (ed.) Jon. S. Wilson, Newnes, (2005).

[204] F. Winquist, P. Wide, T. Eklov, C. Hjort and I. Lundstron, Chrispbread quality evaluation based on fusion of information from the sensor analogies to the human olfaction, auditory and tactile senses, *International Journal of Food Process Engineering*, **22**(5), 337–357, (2000).

[205] F. Winquist, P. Wide, I. Lundstrom, An electronic tongue based on volltammetry, *Analytica Chemica Acta*, **357**, 21–31, (1997).

[206] R. C. White-Traut, M. N. Nelson, J. M. Silvestri, U. Vasan, S. Littau, P. Meleedy-Rey, G. Gu and M. Patel, Effect of auditory, tactile, visual, and vestibular intervention on length of stay alertness, and feeding progression in preterm infants, *Developmental Medicine and Child Neurology*, February, **44**(2), 91–97, (2002).

[207] H. Wu, *Sensor Data Fusion for Context-Aware Computing using Dempster-Shafer Theory*, PhD thesis, December, The Robotics Institute, Carnegie Mellon University, Pittsburg, PA, USA, (2003).

[208] S. Yantis, To see is to attend, *Science*, **299**, 54–56, (2003).

[209] E. Zio, P. Baraldi and G. Gola, Feature-based classifier ensembles for diagnosing multiple faults in rotating machinery, *Applied Soft Computing*, **8**(4), 1365–1380, (2008).

Index